Fisheries Management

Fisheries Management

Edited by
Clavin Wilkinson

Larsen & Keller
www.larsen-keller.com

Fisheries Management
Edited by Clavin Wilkinson
ISBN: 978-1-63549-120-3 (Hardback)

© 2017 Larsen & Keller

▤ Larsen & Keller

Published by Larsen and Keller Education,
5 Penn Plaza,
19th Floor,
New York, NY 10001, USA

Cataloging-in-Publication Data

Fisheries management / edited by Clavin Wilkinson.
    p. cm.
Includes bibliographical references and index.
ISBN 978-1-63549-120-3
1. Fishery management. 2. Fisheries. 3. Aquaculture. 4. Fisheries--Climatic factors.
5. Fishery law and legislation. I. Wilkinson, Clavin.
SH328 .F57 2017
338.372--dc23

The publisher's policy is to use permanent paper from mills that operate a sustainable forestry policy. Furthermore, the publisher ensures that the text paper and cover boards used have met acceptable environmental accreditation standards.

Printed and bound in the United States of America.

For more information regarding Larsen and Keller Education and its products, please visit the publisher's website www.larsen-keller.com

Table of Contents

Preface

This book provides comprehensive insights into the field of fisheries management. It talks in detail about the various advancements in this field. Fisheries management refers to the practice of using fisheries science in order to protect and ensure sustainability of fisheries resources. Modern fisheries management incorporates the rules set up by the government and concerned organizations to stop exploitation of these resources. The aim of this text is to impart crucial information to readers about this topic. This textbook attempts to understand the multiple branches that fall under the discipline of fisheries management and how such concepts have practical applications. As the field is emerging at a rapid pace, the contents of the book will help understand the modern concepts and applications of the subject more easily.

To facilitate a deeper understanding of the contents of this book a short introduction of every chapter is written below:

Chapter 1- Fisheries management is a discipline that utilizes the concepts of fisheries science to develop methods that aim to help in the sustainable exploitation of fishery resources. Decisions and planning is integral to this field and is taken with the help of data gathered through constant monitoring and surveillance. This chapter introduces the discipline that is an amalgamation of sustainable fishery with fishery science.

Chapter 2- This chapter examines the fishing industry in a thorough manner by discussing its distinct sectors. This includes topics like commercial fishing, artisanal fishing, fish processing etc. It also demarcates between commercial, traditional and recreational sectors of fishing. The topics discussed in the chapter are of great importance to broaden the existing knowledge on fisheries management.

Chapter 3- This chapter studies the array of techniques that are employed for fishing globally and some of these include longline fishing, cast net, jug fishing, surf fishing and trawling. The topics listed combine simple and basic technologies with modern methods. This chapter discusses the methods of fishing in a critical manner providing key analysis to the subject matter.

Chapter 4- This chapter studies aquaculture or aquafarming, the cultivation of aquatic organisms like fish, crustaceans, water plants etc. It delves into the methods and techniques of the commercial industries of fish farming, oyster farming and shrimp farming. It also has a section dedicated to mariculture, the cultivation of marine organisms in open or enclosed saltwater spaces. The aspects elucidated in this chapter are of vital importance, and provide a better understanding of fisheries.

Chapter 5- The practice of fisheries faces many hindrances and issues that can crop up from overfishing, attack of parasites, unnecessary fish slaughter etc. Conservation and regulation of fish consumption is extremely important to maintain biodiversity of the marine environment. This chapter details these issues with a comprehensive analysis of each topic.

Chapter 6- The effects of global warming and pollution have affected the fishing industry as well. Climate change and the acidification of oceans are strikingly changing aquatic ecosystems. The rise in sea-levels directly affect communities and groups that practice fishing and aquaculture. This chapter focuses on the effects of climate change, its impact on the oceans and other water bodies and also on fish stocks.

Chapter 7- The term population dynamics in fisheries refers to the growth or fall in the population of aquatic organisms. It forms a basis for understanding the issues of habitat destruction, optimal harvesting rates and predation. Fisheries depend on effective analysis of data gathered through national and international efforts. This chapter deals exclusively with the tools and theories used in measuring population dynamics and understanding viable and sustainable fishing techniques.

Chapter 8- Understanding the complexities of the fishing industry and that disputes can arise in international waters, several laws and regulations are in place to allow judicious fishing while conserving ecosystems. This chapter provides knowledge about seafood safety regulations, individual fishing quotas while also tackling issues regarding the cultivation of genetically modified organisms (GMOs). Fisheries management is best understood in confluence with the major topics listed in the following chapter.

I owe the completion of this book to the never-ending support of my family, who supported me throughout the project.

Editor

Introduction to Fisheries Management

Fisheries management is a discipline that utilizes the concepts of fisheries science to develop methods that aim to help in the sustainable exploitation of fishery resources. Decisions and planning is integral to this field and is taken with the help of data gathered through constant monitoring and surveillance. This chapter introduces the discipline that is an amalgamation of sustainable fishery with fishery science.

Fisheries Management

Fisheries management draws on fisheries science in order to find ways to protect fishery resources so sustainable exploitation is possible. Modern fisheries management is often referred to as a governmental system of appropriate management rules based on defined objectives and a mix of management means to implement the rules, which are put in place by a system of monitoring control and surveillance. According to the Food and Agriculture Organization of the United Nations (FAO), there are "no clear and generally accepted definitions of fisheries management". However, the working definition used by the FAO and much cited elsewhere is:

The integrated process of information gathering, analysis, planning, consultation, decision-making, allocation of resources and formulation and implementation, with enforcement as necessary, of regulations or rules which govern fisheries activities in order to ensure the continued productivity of the resources and the accomplishment of other fisheries objectives.

History

Fisheries have been explicitly managed in some places for hundreds of years. More than 80 percent of the worlds commercial exploitation of fish and shellfish are harvest from natural occurring populations in the oceans and freshwater areas. For example, the Māori people, New Zealand residents for about 700 years, had prohibitions against taking more than what could be eaten and about giving back the first fish caught as an offering to sea god Tangaroa. Starting in the 18th century attempts were made to regulate fishing in the North Norwegian fishery. This resulted in the enactment of a law in 1816 on the Lofoten fishery, which established in some measure what has come to be known as territorial use rights.

"The fishing banks were divided into areas belonging to the nearest fishing base on land and further subdivided into fields where the boats were allowed to fish. The allocation of the fishing fields was in the hands of local governing committees, usually headed by the owner of the onshore facilities which the fishermen had to rent for accommodation and for drying the fish."

Governmental resource protection-based fisheries management is a relatively new idea, first developed for North European fisheries after the first Overfishing Conference held in London in

1936. In 1957 British fisheries researchers Ray Beverton and Sidney Holt published a seminal work on North Sea commercial fisheries dynamics. In the 1960s the work became the theoretical platform for North European management schemes.

After some years away from the field of fisheries management, Beverton criticized his earlier work in a paper given at the first World Fisheries Congress in Athens in 1992. "The Dynamics of Exploited Fish Populations" expressed his concerns, including the way his and Sidney Holt's work had been misinterpreted and misused by fishery biologists and managers during the previous 30 years. Nevertheless, the institutional foundation for modern fishery management had been laid.

A report by Prince Charles' International Sustainability Unit, the New York-based Environmental Defense Fund and 50in10 published in July 2014 estimated global fisheries were adding $270 billion a year to global GDP, but by full implementation of sustainable fishing, that figure could rise by an extra amount of as much as $50 billion.

Political Objectives

According to the FAO, fisheries management should be based explicitly on political objectives, ideally with transparent priorities. Typical political objectives when exploiting a fish resource are to:

- maximize sustainable biomass yield
- maximize sustainable economic yield
- secure and increase employment
- secure protein production and food supplies
- increase export income

Such political goals can also be a weak part of fisheries management, since the objectives can conflict with each other.

International Objectives

Fisheries objectives need to be expressed in concrete management rules. In most countries fisheries management rules should be based on the internationally agreed, though non-binding, Code of Conduct for Responsible Fisheries, agreed at a meeting of the U.N.'s Food and Agriculture Organization FAO session in 1995. The precautionary approach it prescribes is typically implemented in concrete management rules as minimum spawning biomass, maximum fishing mortality rates, etc. In 2005 the UBC Fisheries Centre at the University of British Columbia comprehensively reviewed the performance of the world's major fishing nations against the Code.

International agreements are required in order to regulate fisheries in international waters. The desire for agreement on this and other maritime issues led to three conferences on the Law of the Sea, and ultimately to the treaty known as the United Nations Convention on the Law of the Sea (UNCLOS). Concepts such as exclusive economic zones (EEZ, extending 200 nautical miles (370 km) from a nation's coasts) allocate certain sovereign rights and responsibilities for resource management to individual countries.

Other situations need additional intergovernmental coordination. For example, in the Mediterranean Sea and other relatively narrow bodies of water, EEZ of 200 nautical miles (370 km) are irrelevant. International waters beyond 12-nautical-mile (22 km) from shore require explicit agreements.

Straddling fish stocks, which migrate through more than one EEZ also present challenges. Here sovereign responsibility must be agreed with neighbouring coastal states and fishing entities. Usually this is done through the medium of a regional organisation set up for the purpose of coordinating the management of that stock.

UNCLOS does not prescribe precisely how fisheries confined only to international waters should be managed. Several new fisheries (such as high seas bottom trawling fisheries) are not (yet) subject to international agreement across their entire range. In November 2004 the UN General Assembly issued a resolution on Fisheries that prepared for further development of international fisheries management law.

Management Mechanisms

Many countries have set up Ministries/Government Departments, named "Ministry of Fisheries" or similar, controlling aspects of fisheries within their exclusive economic zones. Four categories of management means have been devised, regulating either input/investment, or output, and operating either directly or indirectly:

	Inputs	**Outputs**
Indirect	Vessel licensing	Catching techniques
Direct	Limited entry	Catch quota and technical regulation

Technical means may include:

- prohibiting devices such as bows and arrows, and spears, or firearms
- prohibiting nets
- setting minimum mesh sizes
- limiting the average potential catch of a vessel in the fleet (vessel and crew size, gear, electronic gear and other physical "inputs".
- prohibiting bait
- snagging
- limits on fish traps
- limiting the number of poles or lines per fisherman
- restricting the number of simultaneous fishing vessels
- limiting a vessel's average operational intensity per unit time at sea
- limiting average time at sea

Catch Quotas

Systems that use *individual transferable quotas* (ITQ), also called individual fishing quota limit the total catch and allocate shares of that quota among the fishers who work that fishery. Fishers can buy/sell/trade shares as they choose.

A large scale study in 2008 provided strong evidence that ITQ's can help to prevent fishery collapse and even restore fisheries that appear to be in decline. Other studies have shown negative socio-economic consequences of ITQs, especially on small-sclale fisheries. These consequences include concentration of quota in that hands of few fishers; increased number of inactive fishers leasing their quotas to others (a phenomenon known as armchair fishermen); and detrimental effects on coastal communities.

Precautionary Principle

The *Fishery Manager's Guidebook* issued in 2009 by the FAO of the United Nations, advises that the precautionary approach or principle should be applied when "ecosystem resilience and human impact (including reversibility) are difficult to forecast and hard to distinguish from natural changes." The precautionary principle suggests that when an action risks harm, it should not be proceeded with until it can be scientifically proven to be safe. Historically fishery managers have applied this principle the other way round; fishing activities have not been curtailed until it has been proven that they have already damaged existing ecosystems. In a paper published in 2007, Shertzer and Prager suggested that there can be significant benefits to stock biomass and fishery yield if management is stricter and more prompt.

Fisheries Law

Fisheries law is an emerging and specialized area of law which includes the study and analysis of different fisheries management approaches, including seafood safety regulations and aquaculture regulations. Despite its importance, this area is rarely taught at law schools around the world, which leaves a vacuum of advocacy and research.

Climate Change

In the past, changing climate has affected inland and offshore fisheries and such changes are likely to continue. From a fisheries perspective, the specific driving factors of climate change include rising water temperature, alterations in the hydrologic cycle, changes in nutrient fluxes, and relocation of spawning and nursery habitat. Further, changes in such factors would affect resources at all levels of biological organization, including the genetic, organism, population, and ecosystem levels.

Population Dynamics

Population dynamics describes the growth and decline of a given fishery stock over time, as controlled by birth, death and migration. It is the basis for understanding changing fishery patterns and issues such as habitat destruction, predation and optimal harvesting rates. The population dynamics of fisheries has been traditionally used by fisheries scientists to determine sustainable yields.

The basic accounting relation for population dynamics is the BIDE model:

$$N_1 = N_0 + B - D + I - E$$

where N_1 is the number of individuals at time 1, N_0 is the number of individuals at time 0, B is the number of individuals born, D the number that died, I the number that immigrated, and E the number that emigrated between time 0 and time 1. While immigration and emigration can be present in wild fisheries, they are usually not measured.

Care is needed when applying population dynamics to real world fisheries. In the past, over-simplistic modelling, such as ignoring the size, age and reproductive status of the fish, focusing solely on a single species, ignoring bycatch and physical damage to the ecosystem, has accelerated the collapse of key stocks.

Ecosystem Based Fisheries

> *We propose that rebuilding ecosystems, and not sustainability per se, should be the goal of fishery management. Sustainability is a deceptive goal because human harvesting of fish leads to a progressive simplification of ecosystems in favour of smaller, high turnover, lower trophic level fish species that are adapted to withstand disturbance and habitat degradation.*
>
> — *Tony Pitcher and Daniel Pauly,*

According to marine ecologist Chris Frid, the fishing industry points to pollution and global warming as the causes of unprecedentedly low fish stocks in recent years, writing, "Everybody would like to see the rebuilding of fish stocks and this can only be achieved if we understand all of the influences, human and natural, on fish dynamics." Overfishing has also had an effect. Frid adds, "Fish communities can be altered in a number of ways, for example they can decrease if particular sized individuals of a species are targeted, as this affects predator and prey dynamics. Fishing, however, is not the sole perpetrator of changes to marine life - pollution is another example [...] No one factor operates in isolation and components of the ecosystem respond differently to each individual factor."

In contrast to the traditional approach of focusing on a single species, the ecosystem-based approach is organized in terms of ecosystem services. Ecosystem-based fishery concepts have been implemented in some regions. In 2007 a group of scientists offered the following *ten commandments*

- Keep a perspective that is holistic, risk-adverse and adaptive.

- Maintain an "old growth" structure in fish populations, since big, old and fat female fish have been shown to be the best spawners, but are also susceptible to overfishing.

- Characterize and maintain the natural spatial structure of fish stocks, so that management boundaries match natural boundaries in the sea.

- Monitor and maintain seafloor habitats to make sure fish have food and shelter.

- Maintain resilient ecosystems that are able to withstand occasional shocks.

- Identify and maintain critical food-web connections, including predators and forage species.

- Adapt to ecosystem changes through time, both short-term and on longer cycles of decades or centuries, including global climate change.

- Account for evolutionary changes caused by fishing, which tends to remove large, older fish.

- Include the actions of humans and their social and economic systems in all ecological equations.

- Report to Congress (2009): The State of Science to Support an Ecosystem Approach to Regional Fishery Management National Marine Fisheries Service, NOAA Technical Memorandum NMFS-F/SPO-96.

Elderly Maternal Fish

Old fat female rockfish are the best producers

Traditional management practices aim to reduce the number of old, slow-growing fish, leaving more room and resources for younger, faster-growing fish. Most marine fish produce huge numbers of eggs. The assumption was that younger spawners would produce plenty of viable larvae.

However, 2005 research on rockfish shows that large, elderly females are far more important than younger fish in maintaining productive fisheries. The larvae produced by these older maternal fish grow faster, survive starvation better, and are much more likely to survive than the offspring of younger fish. Failure to account for the role of older fish may help explain recent collapses of some major US West Coast fisheries. Recovery of some stocks is expected to take decades. One way to prevent such collapses is to establish marine reserves, where fishing is not allowed and fish populations age naturally.

Data Quality

According to fisheries scientist Milo Adkison, the primary limitation in fisheries management decisions is the absence of quality data. Fisheries management decisions are often based on population models, but the models need quality data to be effective. He asserts that scientists and fishery managers would be better served with simpler models and improved data.

The most reliable source for summary statistics is the FAO Fisheries Department.

Ecopath

Ecopath, with Ecosim (EwE), is an ecosystem modelling software suite. It was initially a NOAA initiative led by Jeffrey Polovina, later primarily developed at the UBC Fisheries Centre of the University of British Columbia. In 2007, it was named as one of the ten biggest scientific break-throughs in NOAA's 200-year history. The citation states that Ecopath "revolutionized scientists' ability worldwide to understand complex marine ecosystems". Behind this lies two decades of development work by Villy Christensen, Carl Walters, Daniel Pauly, and other fisheries scientists. As of 2010 there are 6000 registered users in 155 countries. Ecopath is widely used in fisheries management as a tool for modelling and visualising the complex relationships that exist in real world marine ecosystems.

Human Factors

Managing fisheries is about managing people and businesses, and not about managing fish. Fish populations are managed by regulating the actions of people. If fisheries management is to be successful, then associated human factors, such as the reactions of fishermen, are of key importance, and need to be understood.

Management regulations must also consider the implications for stakeholders. Commercial fishermen rely on catches to provide for their families just as farmers rely on crops. Commercial fishing can be a traditional trade passed down from generation to generation. Most commercial fishing is based in towns built around the fishing industry; regulation changes can impact an entire town's economy. Cuts in harvest quotas can have adverse effects on the ability of fishermen to compete with the tourism industry.

Performance

The biomass of global fish stocks has been allowed to run down. This biomass is now diminished to the point where it is no longer possible to sustainably catch the amount of fish that could be caught. According to a 2008 UN report, titled *The Sunken Billions: The Economic Justification for Fisheries Reform*, the world's fishing fleets incur a "$US 50 billion annual economic loss" through depleted stocks and poor fisheries management. The report, produced jointly by the World Bank and the UN Food and Agriculture Organization (FAO), asserts that half the world's fishing fleet could be scrapped with no change in catch.

"By improving governance of marine fisheries, society could capture a substantial part of this $50 billion annual economic loss. Through comprehensive reform, the fisheries sector could become a basis for economic growth and the creation of alternative livelihoods in many countries. At the same time, a nation's natural capital in the form of fish stocks could be greatly increased and the negative impacts of the fisheries on the marine environment reduced."

The most prominent failure of fisheries management in recent times has perhaps been the events that lead to the collapse of the northern cod fisheries. More recently, the International Consortium of Investigative Journalists produced a series of journalistic investigations called *Looting the seas*. These detail investigations into the black market for bluefin tuna, the subsidies propping up the Spanish fishing industry, and the overfishing of the Chilean jack mackerel.

Fisheries Science

Fisheries science is typically taught in a university setting, and can be the focus of an undergraduate, master's or Ph.D. program. Some universities offer fully integrated programs in fisheries science.

Fisheries Research

Fisheries research vessels (FRVs) require platforms which are capable of towing different types of fishing nets, collecting plankton or water samples from a range of depths, and carrying acoustic fish-finding equipment. Fisheries research vessels are often designed and built along the same lines as a large fishing vessel, but with space given over to laboratories and equipment storage, as opposed to storage of the catch.

Notable Contributors

Members of this list meet one or more of the following criteria: 1) Author of widely cited peer-reviewed articles on fisheries, 2) Author of major reference work in fisheries, 3) Founder of major fisheries journal, museum or other related organisation 4) Person most notable for other reasons who has also worked in fisheries science.

Ransom A. Myers

Daniel Pauly

Ray Hilborn

Contributor	Nationality	Born	Died	Contribution
Baird, Spencer F	American	1823	1887	Founding scientist of the United States Fish Commission
Baranov, Fedor I	Russian	1886	1965	Baranov has been called the grandfather of fisheries population dynamics. The Baranov catch equation of 1918 is perhaps the most used equation in fisheries modelling.

Beverton, Ray	English	1922	1985	Fisheries biologist known for the Beverton–Holt model (with Sidney Holt), credited with being one of the founders of fisheries science
Christensen, Villy	Danish		-	Fisheries scientist and ecosystem modeller, known for his work on the development of Ecopath
Cobb, John N	American	1868	1930	Founder of the first college of fisheries in the United States, the University of Washington College of Fisheries, in 1919
Cooke, Steven J	Canadian	1974		Academic known for contributions to recreational fisheries science, inland fisheries and Conservation Physiology
Cushing, David	English	1920	2008	Fisheries biologist, who is credited with the development of the match/mismatch hypothesis
Everhart, W Harry	American	1918	1994	Fisheries scientist, educator, administrator and author of several widely used fisheries texts
Froese, Rainer	German	1950	-	Known for his work on the development and coordination of FishBase
Green, Seth	American	1817	1888	Pioneer in fish farming who established the first fish hatchery in the United States
Halver, John	American	1922	2012	His pioneering work on the nutritional needs of fish led to modern methods of fish farming and fish feed production. He has been called the father of fish nutrition.

Hempel, Gotthilf	German	1929	-	Marine biologist and oceanographer, and co-founder of the Alfred Wegener Institute for Polar and Marine Research
Herwig, Walther	German	1838	1912	Lawyer and promoter of high seas fishing and research
Hilborn, Ray	Canadian	1947	-	Fisheries biologist with strong contributions in fisheries management
Hjort, Johan	Norwegian	1869	1948	Fisheries biologist, marine zoologist and oceanographer
Hofer, Bruno	German	1861	1916	Fishery scientist credited with being the founder of fish pathology
Holt, Sidney	English	1926	-	Fisheries biologist known for the Beverton–Holt model (with Ray Beverton), credited with being one of the founders of fisheries science
Kils, Uwe	German		-	Marine biologist specializing in planktology. Inventor of the ecoSCOPE
Lackey, Robert T	Canadian	1944	-	Fisheries scientist and political scientist known for his work involving the role of science in policy making
Margolis, Leo	Canadian	1927	1997	Parasitologist and head of the Pacific Biological Station in Nanaimo, British Columbia
McKay, R J	Australian			Biologist and a specialist in translocated freshwater fishes
Myers, Ransom A	Canadian	1952	2007	Marine biologist and conservationist
Pauly, Daniel	French	1946		Prominent fisheries scientist, known for his work studying human impacts on global fisheries

Pitcher, Tony J			-	Known for work on the impacts of fishing, management appraisals and the shoaling behavior of fish
Rice, Michael A	American	1955	-	Known for work on molluscan fisheries
Ricker, Bill	Canadian	1908	2001	Fisheries biologist, known for the Ricker model, credited with being one of the founders of fisheries science
Ricketts, Ed	American	1897	1948	A colourful marine biologist and philosopher who introduced ecology to fisheries science.
Roberts, Callum			-	Marine conservation biologist, known for his work on the role marine reserves play in protecting marine ecosystems
Rosenthal, Harald	German	1937	-	Hydrobiologist known for his work in fish farming and ecology
Safina, Carl	American	1955	-	Author of several writings on marine ecology and the ocean
Sars, Georg Ossian	Norwegian	1837	1927	Marine biologist credited with the discovery of a number of new species and known for his analysis of cod fisheries
Schaefer, Milner Baily	American	1912	1970	Notable for work on the population dynamics of fisheries
Schweder, Tore	Norwegian	1943	-	Statistician whose work includes the assessment of marine resources
Sumaila, Ussif Rashid	Nigerian		-	Notable for his analysis of the economic aspects of fisheries
Utter, Fred M	American	1931	-	Notable as the founding father of the field of fishery genetics and his influence on marine conservation

| von Bertalanffy, Ludwig | Austrian | 1901 | 1972 | Biologist and founder of general systems theory |
| Walters, Carl | American | | - | Biologist known for his work involving fisheries stock assessments, the adaptive management concept, and ecosystem modeling |

Fisheries scientists sorting a catch of small fish and Norway lobster

Sustainable Fishery

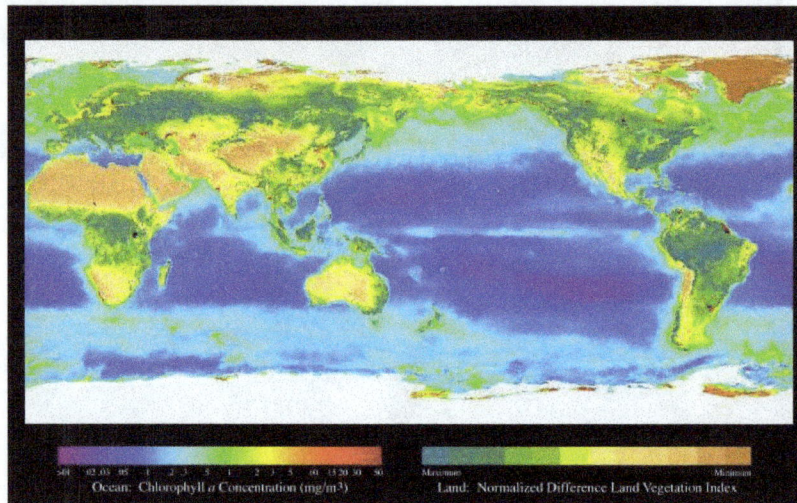

SeaWiFS map showing the levels of primary production in the world's oceans

A conventional idea of a sustainable fishery is that it is one that is harvested at a sustainable rate, where the fish population does not decline over time because of fishing practices. Sustainability in fisheries combines theoretical disciplines, such as the population dynamics of fisheries, with practical strategies, such as avoiding overfishing through techniques such as individual fishing quotas, curtailing destructive and illegal fishing practices by lobbying for appropriate law and policy, setting up protected areas, restoring collapsed fisheries, incorporating all externalities involved in harvesting marine ecosystems into fishery economics, educating stakeholders and the wider public, and developing independent certification programs.

Some primary concerns around sustainability are that heavy fishing pressures, such as overexploitation and growth or recruitment overfishing, will result in the loss of significant potential yield; that stock structure will erode to the point where it loses diversity and resilience to environmental fluctuations; that ecosystems and their economic infrastructures will cycle between collapse and recovery; with each cycle less productive than its predecessor; and that changes will occur in the trophic balance (fishing down marine food webs).

Overview

> *Sustainable management of fisheries cannot be achieved without an acceptance that the long-term goals of fisheries management are the same as those of environmental conservation*

— *Daniel Pauly* and Dave Preikshot,

Global wild fisheries are believed to have peaked and begun a decline, with valuable habitats, such as estuaries and coral reefs, in critical condition. Current aquaculture or farming of piscivorous fish, such as salmon, does not solve the problem because farmed piscivores are fed products from wild fish, such as forage fish. Salmon farming also has major negative impacts on wild salmon. Fish that occupy the higher trophic levels are less efficient sources of food energy.

Fishery ecosystems are an important subset of the wider marine environment. This article documents the views of fisheries scientists and marine conservationists about innovative approaches towards sustainable fisheries.

History

> *In the end, we will conserve only what we love; we will love only what we understand; and we will understand only what we are taught*

— Senegalese conservationist *Baba Dioum,*

In his 1883 inaugural address to the International Fisheries Exhibition in London, Thomas Huxley asserted that overfishing or "permanent exhaustion" was scientifically impossible, and stated that probably "all the great sea fisheries are inexhaustible". In reality, by 1883 marine fisheries were already collapsing. The United States Fish Commission was established 12 years earlier for the purpose of finding why fisheries in New England were declining. At the time of Huxley's address, the Atlantic halibut fishery had already collapsed (and has never recovered).

Traditional Management of Fisheries

Traditionally, fisheries management and the science underpinning it was distorted by its "narrow focus on target populations and the corresponding failure to account for ecosystem effects leading to declines of species abundance and diversity" and by perceiving the fishing industry as "the sole legitimate user, in effect the owner, of marine living resources." Historically, stock assessment scientists usually worked in government laboratories and considered their work to be providing services to the fishing industry. These scientists dismissed conservation issues and distanced themselves from the scientists and the science that raised the issues. This happened even as commercial fish stocks deteriorated, and even though many governments were signatories to binding conservation agreements.

Defining Sustainability

The notion of sustainable development is sometimes regarded as an unattainable, even illogical notion because development inevitably depletes and degrades the environment.

Ray Hilborn, of the University of Washington, distinguishes three ways of defining a sustainable fishery:

- *Long term constant yield* is the idea that undisturbed nature establishes a steady state that changes little over time. Properly done, fishing at up to maximum sustainable yield allows nature to adjust to a new steady state, without compromising future harvests. However, this view is naive, because constancy is not an attribute of marine ecosystems, which dooms this approach. Stock abundance fluctuates naturally, changing the potential yield over short and long term periods.

- *Preserving intergenerational equity* acknowledges natural fluctuations and regards as unsustainable only practices which damage the genetic structure destroy habitat, or deplete stock levels to the point where rebuilding requires more than a single generation. Providing rebuilding takes only one generation, overfishing may be economically foolish, but it is not unsustainable. This definition is widely accepted.

- *Maintaining a biological, social and economic system* considers the health of the human ecosystem as well as the marine ecosystem. A fishery which rotates among multiple species can deplete individual stocks and still be sustainable so long as the ecosystem retains its intrinsic integrity. Such a definition might consider as sustainable fishing practices that lead to the reduction and possible extinction of some species.

Social Sustainability

Fisheries and aquaculture are, directly or indirectly, a source of livelihood for over 500 million people, mostly in developing countries.

Social sustainability can conflict with biodiversity. A fishery is socially sustainable if the fishery ecosystem maintains the ability to deliver products the society can use. Major species shifts within the ecosystem could be acceptable as long as the flow of such products continues. Humans have been operating such regimes for thousands of years, transforming many ecosystems, depleting or driving to extinction many species.

66
> *To a great extent, sustainability is like good art, it is hard to describe but we know it when we see it.*
99

— *Ray Hilborn,*

According to Hilborn, the "loss of some species, and indeed transformation of the ecosystem is not incompatible with sustainable harvests." For example, in recent years, barndoor skates have been caught as bycatch in the western Atlantic. Their numbers have severely declined and they will probably go extinct if these catch rates continue. Even if the barndoor skate goes extinct, changing the ecosystem, there could still be sustainable fishing of other commercial species.

Reconciling Fisheries with Conservation

Management goals might consider the impact of salmon on bear and river ecosystems

At the Fourth World Fisheries Congress in 2004, Daniel Pauly asked, "How can fisheries science and conservation biology achieve a reconciliation?", then answered his own question, "By accepting each other's essentials: that fishing should remain a viable occupation; and that aquatic ecosystems and their biodiversity are allowed to persist."

A relatively new concept is relationship farming. This is a way of operating farms so they restore the food chain in their area. Re-establishing a healthy food chain can result in the farm automatically filtering out impurities from feed water and air, feeding its own food chain, and additionally producing high net yields for harvesting. An example is the large cattle ranch Veta La Palma in southern Spain. Relationship farming was first made popular by Joel Salatin who created a 220 hectare relationship farm featured prominently in Michael Pollan's book *The Omnivore's Dilemma* (2006) and the documentary films, Food, Inc. and Fresh. The basic concept of relationship farming is to put effort into building a healthy food chain, and then the food chain does the hard work.

Obstacles

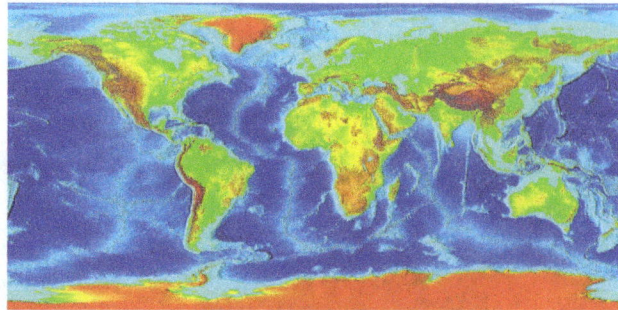

Large areas of the global continental shelf, highlighted in cyan, have had heavy bottom trawls repeatedly dragged over them

Overfishing

Overfishing can be sustainable. According to Hilborn, overfishing can be "a misallocation of societies' resources", but it does not necessarily threaten conservation or sustainability".

Overfishing is traditionally defined as harvesting so many fish that the yield is less than it would be if fishing were reduced. For example, Pacific salmon are usually managed by trying to determine how many spawning salmon, called the "escapement", are needed each generation to produce the maximum harvestable surplus. The optimum escapement is that needed to reach that surplus. If the escapement is half the optimum, then normal fishing looks like overfishing. But this is still sustainable fishing, which could continue indefinitely at its reduced stock numbers and yield. There is a wide range of escapement sizes that present no threat that the stock might collapse or that the stock structure might erode.

On the other hand, overfishing can precede severe stock depletion and fishery collapse. Hilborn points out that continuing to exert fishing pressure while production decreases, stock collapses and the fishery fails, is largely "the product of institutional failure."

Today over 70% of fish species are either fully exploited, overexploited, depleted, or recovering from depletion. If overfishing does not decrease, it is predicted that stocks of all species currently commercially fished for will collapse by 2048."

A Hubbert linearization (Hubbert curve) has been applied to the whaling industry, as well as charting the price of caviar, which depends on sturgeon stocks. Another example is North Sea cod. Comparing fisheries and mineral extraction tells us that human pressure on the environment is causing a wide range of resources to go through a Hubbert depletion cycle.

Island with fringing reef in the Maldives. Coral reefs are dying around the world.

0   100 km   • Aral

Kasachstan

Mo'ynoq

Usbekistan

1960
1970
1980
1990
2000
2010

• Nukus

Shrinking of the Aral Sea

Habitat Modification

Nearly all the world's continental shelves, and large areas of continental slopes, underwater ridges, and seamounts, have had heavy bottom trawls and dredges repeatedly dragged over their surfaces. For fifty years, governments and organizations, such as the Asian Development Bank, have encouraged the fishing industry to develop trawler fleets. Repeated bottom trawling and dredging literally flattens diversity in the benthic habitat, radically changing the associated communities.

Changing The Ecosystem Balance

Since 1950, 90 percent of 25 species of big predator fish have gone.

- How we are emptying our seas *The Sunday Times*, May 10, 2009.

- Pauly, Daniel (2004) Reconciling Fisheries with Conservation: the Challenge of Managing Aquatic Ecosystems Fourth World Fisheries Congress, Vancouver, 2004.

Climate Change

Rising ocean temperatures and ocean acidification are radically altering aquatic ecosystems. Climate change is modifying fish distribution and the productivity of marine and freshwater species. This reduces sustainable catch levels across many habitats, puts pressure on resources needed for aquaculture, on the communities that depend on fisheries, and on the oceans' ability to capture and store carbon (biological pump). Sea level rise puts coastal fishing communities at risk, while changing rainfall patterns and water use impact on inland (freshwater) fisheries and aquaculture.

Ocean Pollution

A recent survey of global ocean health concluded that all parts of the ocean have been impacted by human development and that 41 percent has been fouled with human polluted runoff, overfishing, and other abuses. Pollution is not easy to fix, because pollution sources are so dispersed, and are built into the economic systems we depend on.

The United Nations Environment Programme (UNEP) mapped the impacts of stressors such as climate change, pollution, exotic species, and over-exploitation of resources on the oceans. The report shows at least 75 percent of the world's key fishing grounds may be affected.

Diseases and Toxins

Large predator fish contain significant amounts of mercury, a neurotoxin which can affect fetal development, memory, mental focus, and produce tremors.

Irrigation

Abandoned ship near Aral, Kazakhstan.

Lakes are dependent on the inflow of water from its drainage basin. In some areas, aggressive irrigation has caused this inflow to decrease significantly, causing water depletion and a shrinking of the lake. The most notable example is the Aral Sea, formerly among the four largest lakes in the world, now only a tenth of its former surface area.

Remediation

Fisheries Management

Fisheries management draws on fisheries science to enable sustainable exploitation. Modern fisheries management is often defined as mandatory rules based on concrete objectives and a mix of management techniques, enforced by a monitoring control and surveillance system.

- Ideas and rules: Economist Paul Romer believes sustainable growth is possible providing the right ideas (technology) are combined with the right rules, rather than simply hectoring fishers. There has been no lack of innovative ideas about how to harvest fish. He characterizes failures as primarily failures to apply appropriate rules.

- Fishing subsidies: Government subsidies influence many of the world fisheries. Operating cost subsidies allow European and Asian fishing fleets to fish in distant waters, such as West Africa. Many experts reject fishing subsidies and advocate restructuring incentives globally to help struggling fisheries recover.

- Economics: Another focus of conservationists is on curtailing detrimental human activities by improving fisheries' market structure with techniques such as salable fishing quotas, like those set up by the Northwest Atlantic Fisheries Organization, or laws such as those listed below.

- Payment for Ecosystem Services: Environmental Economist, Essam Y Mohammed, argues that by creating direct economic incentives, whereby people are able to receive payment for the services their property provides, will help to establish sustainable fisheries around the world as well as inspire conservation where it otherwise would not.

- Sustainable fisheries certification: A promising direction is the independent certification programs for sustainable fisheries conducted by organizations such as the Marine Stewardship Council and Friend of the Sea. These programs work at raising consumer awareness and insight into the nature of their seafood purchases.

- Ecosystem based fisheries.

Ecosystem Based Fisheries

> *We propose that rebuilding ecosystems, and not sustainability per se, should be the goal of fishery management. Sustainability is a deceptive goal because human harvesting of fish leads to a progressive simplification of ecosystems in favour of smaller, high turnover, lower trophic level fish species that are adapted to withstand disturbance and habitat degradation.*

— *Tony Pitcher* and *Daniel Pauly*,

According to marine ecologist Chris Frid, the fishing industry points to marine pollution and global warming as the causes of recent, unprecedented declines in fish populations. Frid counters that overfishing has also altered the way the ecosystem works. "Everybody would like to see the rebuilding of fish stocks and this can only be achieved if we understand all of the influences, human and natural, on fish dynamics." He adds: "fish communities can be altered in a number of ways, for example they can decrease if particular-sized individuals of a species are targeted, as this affects predator and prey dynamics. Fishing, however, is not the sole cause of changes to marine life—pollution is another example....No one factor operates in isolation and components of the ecosystem respond differently to each individual factor."

The traditional approach to fisheries science and management has been to focus on a single species. This can be contrasted with the ecosystem-based approach. Ecosystem-based fishery concepts have been implemented in some regions. In a 2007 effort to "stimulate much needed discussion" and "clarify the essential components" of ecosystem-based fisheries science, a group of scientists offered the following ten commandments for ecosystem-based fisheries scientists

"

- Keep a perspective that is holistic, risk-adverse and adaptive.

- Maintain an "old growth" structure in fish populations, since big, old and fat female fish have been shown to be the best spawners, but are also susceptible to overfishing.

- Characterize and maintain the natural spatial structure of fish stocks, so that management boundaries match natural boundaries in the sea.

- Monitor and maintain seafloor habitats to make sure fish have food and shelter.

- Maintain resilient ecosystems that are able to withstand occasional shocks.

- Identify and maintain critical food-web connections, including predators and forage species.

- Adapt to ecosystem changes through time, both short-term and on longer cycles of decades or centuries, including global climate change.

- Account for evolutionary changes caused by fishing, which tends to remove large, older fish.

- Include the actions of humans and their social and economic systems in all ecological equations.

"

Marine Protected Areas

Strategies and techniques for marine conservation tend to combine theoretical disciplines, such as population biology, with practical conservation strategies, such as setting up protected areas, as with Marine Protected Areas (MPAs) or Voluntary Marine Conservation Areas. Each nation defines MPAs independently, but they commonly involve increased protection for the area from fishing and other threats.

Marine life is not evenly distributed in the oceans. Most of the really valuable ecosystems are in relatively shallow coastal waters, above or near the continental shelf, where the sunlit waters are often nutrient rich from land runoff or upwellings at the continental edge, allowing photosynthesis, which energizes the lowest trophic levels. In the 1970s, for reasons more to do with oil drilling than with fishing, the U.S. extended its jurisdiction, then 12 miles from the coast, to 200 miles. This made huge shelf areas part of its territory. Other nations followed, extending national control to what became known as the exclusive economic zone (EEZ). This move has had many implications for fisheries conservation, since it means that most of the most productive maritime ecosystems are now under national jurisdictions, opening possibilities for protecting these ecosystems by passing appropriate laws.

Daniel Pauly characterises marine protected areas as "a conservation tool of revolutionary importance that is being incorporated into the fisheries mainstream." The Pew Charitable Trusts have funded various initiatives aimed at encouraging the development of MPAs and other ocean conservation measures.

Fish Farming

There exists concerns that farmed fish cannot produce necessary yields efficiently. For example, farmed salmon eat three pounds of wild fish to produce one pound of salmon.

Laws and Treaties

International laws and treaties related to marine conservation include the 1966 Convention on Fishing and Conservation of Living Resources of the High Seas. United States laws related to marine conservation include the 1972 Marine Mammal Protection Act, as well as the 1972 Marine Protection, Research and Sanctuaries Act which established the National Marine Sanctuaries program. Magnuson-Stevens Fishery Conservation and Management Act.

Awareness Campaigns

Introducing the results of long term monitoring to a local fishermen in Kihnu, Estonia.

Various organizations promote sustainable fishing strategies, educate the public and stakeholders, and lobby for conservation law and policy. The list includes the Marine Conservation Biology Institute and Blue Frontier Campaign in the U.S., The U.K.'s Frontier (the Society for Environmental Exploration) and Marine Conservation Society, Australian Marine Conservation Society, International Council for the Exploration of the Sea (ICES), Langkawi Declaration, Oceana, PROFISH, and the Sea Around Us Project, International Collective in Support of Fishworkers, World Forum of Fish Harvesters and Fish Workers, Frozen at Sea Fillets Association and CEDO.

The United Nations Millennium Development Goals include, as goal #7: target 2, the intention to "reduce biodiversity loss, achieving, by 2010, a significant reduction in the rate of loss", including improving fisheries management to reduce depletion of fish stocks.

Some organizations certify fishing industry players for sustainable or good practices, such as the Marine Stewardship Council and Friend of the Sea.

Other organizations offer advice to members of the public who eat with an eye to sustainability. According to the marine conservation biologist Callum Roberts, four criteria apply when choosing seafood:

- Is the species in trouble in the wild where the animals were caught?

- Does fishing for the species damage ocean habitats?

- Is there a large amount of bycatch taken with the target species?

- Does the fishery have a problem with discards—generally, undersized animals caught and thrown away because their market value is low?

The following organizations have download links for wallet-sized cards, listing good and bad choices:

- Monterey Bay Aquarium Seafood Watch, USA

- Blue Ocean Institute, USA

- Marine Conservation Society, UK

- Australian Marine Conservation Society

- The Southern African Sustainable Seafood Initiative

Data Issues

Data Quality

One of the major impediments to the rational control of marine resources is inadequate data. According to fisheries scientist Milo Adkison (2007), the primary limitation in fisheries management decisions is poor data. Fisheries management decisions are often based on population models, but the models need quality data to be accurate. Scientists and fishery managers would be better served with simpler models and improved data.

Unreported Fishing

Estimates of illegal catch losses range between $10 billion and $23 billion annually, representing between 11 and 26 million tonnes.

- Incidental catch

Shifting Baselines

Shifting baselines is a term which describes the way significant changes to a system are measured against previous baselines, which themselves may represent significant changes from the original state of the system. The term was first used by the fisheries scientist Daniel Pauly in his paper "Anecdotes and the shifting baseline syndrome of fisheries". Pauly developed the term in reference to fisheries management where fisheries scientists sometimes fail to identify the correct "baseline" population size (e.g. how abundant a fish species population was *before* human exploitation) and thus work with a shifted baseline. He describes the way that radically depleted fisheries were evaluated by experts who used the state of the fishery at the start of their careers as the baseline, rather than the fishery in its untouched state. Areas that swarmed with a particular species hundreds of years ago, may have experienced long term decline, but it is the level of decades previously that is considered the appropriate reference point for current populations. In this way large declines in ecosystems or species over long periods of time were, and are, masked. There is a loss of perception of change that occurs when each generation redefines what is "natural".

Looting the seas

Looting the seas is the name given by the International Consortium of Investigative Journalists to a series of journalistic investigations into areas directly affecting the sustainability of fisheries. So far they have investigated three areas involving fraud, negligence and overfishing:

- The black market in bluefin tuna

- Subsidies propping up the Spanish fishing industry

- Overfishing of the southern jack mackerel

References

- Arnason, R; Kelleher, K; Willmann, R (2008). The Sunken Billions: The Economic Justification for Fisheries Reform. World Bank and FAO. ISBN 978-0-8213-7790-1.

- Beverton, R. J. H.; Holt, S. J. (1957). On the Dynamics of Exploited Fish Populations. Fishery Investigations Series II Volume XIX. Chapman and Hall (Blackburn Press, 2004). ISBN 978-1-930665-94-1.

- Caddy JF and Mahon R (1995) "Reference points for fisheries management" FAO Fisheries technical paper 347, Rome. ISBN 92-5-103733-7

- McGoodwin JR (2001) Understanding the cultures of fishing communities. A key to fisheries management and food security FAO Fisheries, Technical Paper 401. ISBN 978-92-5-104606-7.

- Morgan, Gary; Staples, Derek and Funge-Smith, Simon (2007) Fishing capacity management and illegal, unreported and unregulated fishing in Asia FAO RAP Publication. 2007/17. ISBN 978-92-5-005669-2

- Townsend, R; Shotton, Ross and Uchida, H (2008) Case studies in fisheries self-governance FAO Fisheries Technical Paper. No 504. ISBN 978-92-5-105897-8

- Walters, Carl J. and Steven J. D. Martell (2004) Fisheries ecology and management Princeton University Press. ISBN 978-0-691-11545-0.

- Hart, Paul J B and Reynolds, John D (2002) Handbook of Fish Biology and Fisheries, Chapter 1, The human dimensions of fisheries science. Blackwell Publishing. ISBN 0-632-06482-X

- Megrey BA and Moksness E (eds) (2009) Computers in Fisheries Research second edition, Springer. ISBN 978-1-4020-8635-9. doi:10.1007/978-1-4020-8636-6_1

- Payne A, Cotter AJR, Cotter J and Potter T (2008) Advances in fisheries science: 50 years on from Beverton and Holt John Wiley and Sons. ISBN 978-1-4051-7083-3.

- Norse, Elliott A. and Crowder, Larry B. (Eds.) (2005) Marine Conservation Biology: The Science of Maintaining the Sea's Biodiversity, Island Press. ISBN 978-1-55963-662-9

- McLeod, Karen and Leslie. Heather (Eds.) (2009) Ecosystem-Based Management for the Oceans Island Press. ISBN 978-1-59726-155-5

- Berkes F, Mahon R, McConney P, Pollnac R and Pomeroy R (2001) Managing Small-Scale Fisheries: Alternative Directions and Methods IDRC,. ISBN 978-0-88936-943-6

- Mann, Kenneth and Lazier, John (3rd Ed. 2005) Dynamics of Marine Ecosystems: Biological-Physical Interactions in the Oceans Wiley-Blackwell. ISBN 978-1-4051-1118-8

- Norse EA and Crowder LB (Eds) (2005) Marine conservation biology: the science of maintaining the sea's biodiversity Island Press. ISBN 978-1-55963-662-9

Fishing Industry: An Overview

This chapter examines the fishing industry in a thorough manner by discussing its distinct sectors. This includes topics like commercial fishing, artisanal fishing, fish processing etc. It also demarcates between commercial, traditional and recreational sectors of fishing. The topics discussed in the chapter are of great importance to broaden the existing knowledge on fisheries management.

Fishing Industry

Global production of aquatic organisms in million tonnes, since 1950, as reported by the FAO

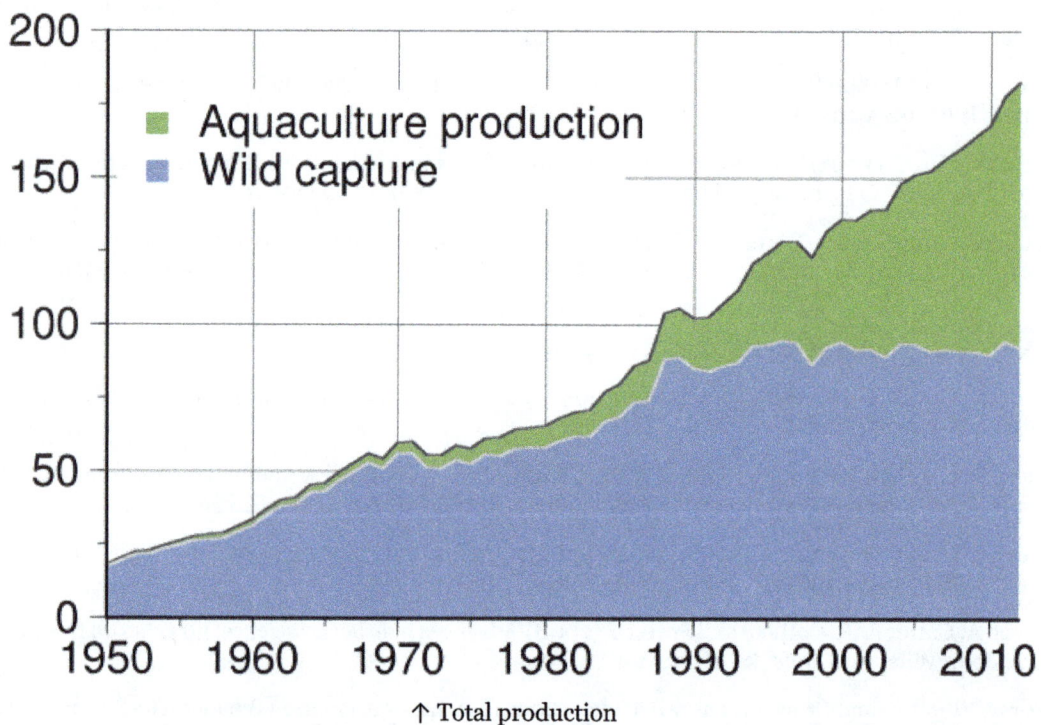

↑ Total production

The fishing industry includes any industry or activity concerned with taking, culturing, processing, preserving, storing, transporting, marketing or selling fish or fish products. It is defined by the Food and Agriculture Organization as including recreational, subsistence and commercial fishing, and the harvesting, processing, and marketing sectors. The commercial activity is aimed at the delivery of fish and other seafood products for human consumption or as input factors in other industrial processes. Directly or indirectly, the livelihood of over 500 million people in developing countries depends on fisheries and aquaculture.

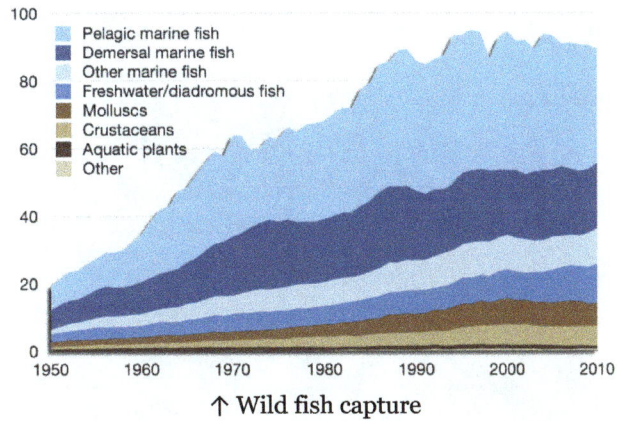

↑ Wild fish capture

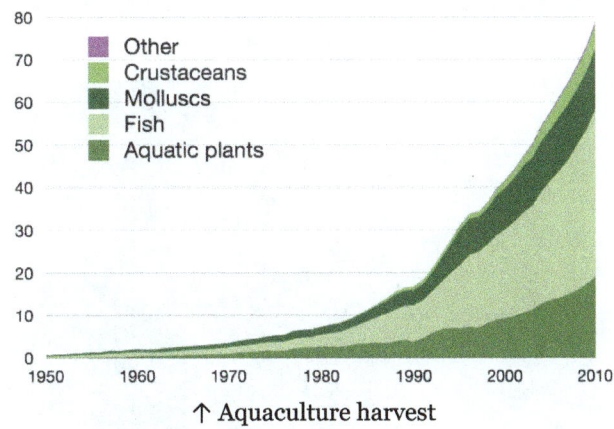

↑ Aquaculture harvest

Sectors

Using a special tuna knife at Tsukiji fish market in Tokyo

Sea urchin roe

Fresh seafood laid out on one of several floating barge vendors

There are three principal industry sectors:

- The commercial sector: comprises enterprises and individuals associated with wild-catch or aquaculture resources and the various transformations of those resources into products for sale. It is also referred to as the "seafood industry", although non-food items such as pearls are included among its products.

- The traditional sector: comprises enterprises and individuals associated with fisheries resources from which aboriginal people derive products in accordance with their traditions.

- The recreational sector: comprises enterprises and individuals associated for the purpose of recreation, sport or sustenance with fisheries resources from which products are derived that are not for sale.

Commercial Sector

The commercial sector of the fishing industry comprises the following chain:

1. Commercial fishing and fish farming which produce the fish

2. Fish processing which produce the fish products

3. Marketing of the fish products

World Production

Fish are harvested by commercial fishing and aquaculture.

According to the Food and Agriculture Organization (FAO), the world harvest in 2005 consisted of 93.3 million tonnes captured by commercial fishing in wild fisheries, plus 48.1 million tonnes produced by fish farms. In addition, 1.3 million tons of aquatic plants (seaweed etc.) were captured in wild fisheries and 14.8 million tons were produced by aquaculture. The number of individual fish caught in the wild has been estimated at 0.97-2.7 trillion per year (not counting fish farms or marine invertebrates).

Following is a table of the 2011 world fishing industry harvest in tonnes by capture and by aquaculture.

	Capture (ton)	Aquaculture (ton)	Total (ton)
Total	94,574,113	83,729,313	178,303,426
Aquatic plant	1,085,143	20,975,361	22,060,504
Aquatic animal	93,488,970	62,753,952	156,202,922

Commercial Fishing

The top producing countries were, in order, the People's Republic of China (excluding Hong Kong and Taiwan), Peru, Japan, the United States, Chile, Indonesia, Russia, India, Thailand, Norway and Iceland. Those countries accounted for more than half of the world's production; China alone accounted for a third of the world's production.

Fish Farming

Aquaculture is the cultivation of aquatic organisms. Unlike fishing, aquaculture, also known as aquafarming, is the cultivation of aquatic populations under controlled conditions. Mariculture refers to aquaculture practiced in marine environments. Particular kinds of aquaculture include algaculture (the production of kelp/seaweed and other algae); fish farming; shrimp farming, shellfish farming, and the growing of cultured pearls.

Fish farming involves raising fish commercially in tanks or enclosed pools, usually for food. Fish species raised by fish farms include carp, salmon, tilapia, catfish and cod. Increasing demands on wild fisheries by commercial fishing operations have caused widespread overfishing. Fish farming offers an alternative solution to the increasing market demand for fish and fish protein.

Fish Processing

Fish processing is the processing of fish delivered by commercial fisheries and fish farms. The larger fish processing companies have their own fishing fleets and independent fisheries. The products of the industry are usually sold wholesale to grocery chains or to intermediaries.

Fish processing can be subdivided into two categories: fish handling (the initial processing of raw fish) and fish products manufacturing. Aspects of fish processing occur on fishing vessels, fish processing vessels, and at fish processing plants.

Another natural subdivision is into primary processing involved in the filleting and freezing of fresh fish for onward distribution to fresh fish retail and catering outlets, and the secondary processing that produces chilled, frozen and canned products for the retail and catering trades.

Fish Products

Fisheries are estimated to currently provide 16% of the world population's protein. The flesh of many fish are primarily valued as a source of food; there are many edible species of fish. Other marine life taken as food includes shellfish, crustaceans, sea cucumber, jellyfish and roe.

Fish and other marine life are also be used for many other uses: pearls and mother-of-pearl, sharkskin and rayskin. Sea horses, star fish, sea urchins and sea cucumber are used in traditional Chinese medicine. Tyrian purple is a pigment made from marine snails, sepia is a pigment made from the inky secretions of cuttlefish. Fish glue has long been valued for its use in all manner of products. Isinglass is used for the clarification of wine and beer. Fish emulsion is a fertilizer emulsion that is produced from the fluid remains of fish processed for fish oil and fish meal.

In the industry the term *seafood products* is often used instead of *fish products*.

Fish Marketing

Fish markets are marketplace used for the trade in and sale of fish and other seafood. They can be dedicated to wholesale trade between fishermen and fish merchants, or to the sale of seafood to individual consumers, or to both. Retail fish markets, a type of wet market, often sell street food as well.

Most shrimp are sold frozen and are marketed in different categories. The live food fish trade is a global system that links fishing communities with markets.

Traditional Sector

The traditional fishing industry, or artisan fishing, are terms used to describe small scale commercial or subsistence fishing practises, particularly using traditional techniques such as rod and tackle, arrows and harpoons, throw nets and drag nets, etc. It does not usually cover the concept of fishing for sport, and might be used when talking about the pressures between large scale modern commercial fishing practises and traditional methods, or when aid programs are targeted specifically at fishing at or near subsistence levels.

Recreational Sector

The recreational fishing industry consists of enterprises such as the manufacture and retailing of fishing tackle and apparel, the payment of license fees to regulatory authorities, fishing books and magazines, the design and building of recreational fishing boats, and the provision of accommodation, fishing boats for charter, and guided fishing adventures.

Lift nets in Cà Mau, Vietnam

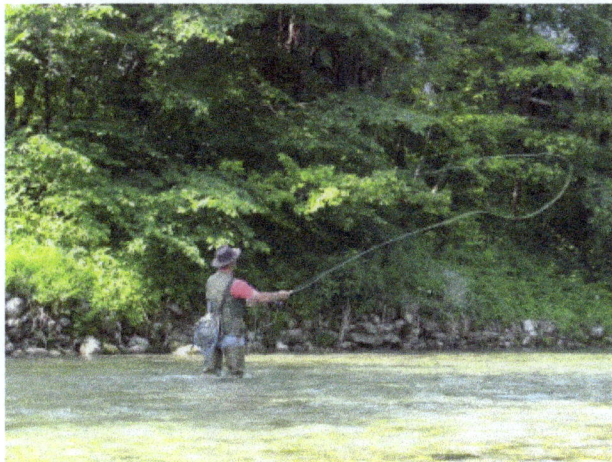

Fly fishing in a river

International Problems

The ocean covers 71% of the earth's surface and 80% of the value of exploited marine resources are attributed to the fishing industry. The fishing industry has provoked various international disputes as wild fish capture rose to a peak about the turn of the century, and has since started a gradual decline. Iceland, Japan, and Portugal are the greatest consumers of seafood per capita in the world.

Problems in The Americas

Chile and Peru are countries with high fish consumption, and therefore had troubles regarding fish industry. In 1947, Chile and Peru first adopted the 200 nautical miles of Exclusive Economic Zone for their shore, and in 1982, UN formally adopted this term. In 2000s, Chile and Peru suffered serious fish crisis because of excessive fishing and lack of proper regulations, and now political power play in the area is rekindled. From the late 1950s, offshore bottom trawlers began exploiting the deeper part, leading to a large catch increase and a strong decline in the underlying biomass. The stock collapsed to extremely low levels in the early 1990s and this is a well-known example of non-excludable, non-rivalrous public good in economics, causing free-rider problems.

Problems in Europe

Iceland is one of the largest consumers in the world and in 1972, a dispute occurred between UK and Iceland because of Iceland's announcement of Exclusive Economic Zone (EEZ) to reduce overfishing. This dispute is called the Cod War, direct confrontations between Icelandic patrol vessels and British warships. Nowadays in Europe in general, countries are searching for a way to recover fishing industry. Overfishing of EU fisheries is costing 3.2 billion euros a year and 100,000 jobs according to a report. So Europe is constantly looking for some collective actions to prevent overfishing.

Problems in Asia

Japan, China and Korea are some of the greatest consumers of fish, and have some disputes over Exclusive Economic Zone. In 2011, due to a serious earthquake, the nuclear power facility in Fukushima was damaged. Ever since, huge amount of contaminated water leaked and is entering the oceans. Tokyo Electric Power Company (Tepco) admitted that around 300 tonnes of highly radioactive water had leaked from a storage tank on the site. In the Kuroshio Current, the sea near Fukushima, about 11 countries catch fish. Not only the surrounding countries such as Japan, Korea and China, but also the countries like Ukraine, Spain and Russia have boats in the Kuroshio Current. In September 2013, South Korea banned all fish imports from eight Japanese prefectures, concerning radioactive water leak from the Fukushima nuclear plant.

Vessel Monitoring System

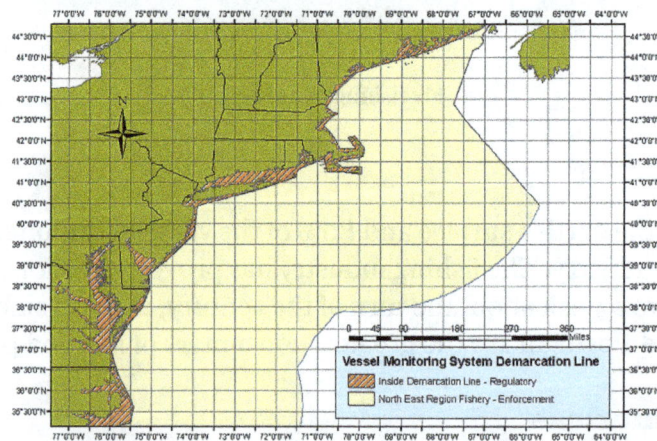

NOAA Atlantic fisheries area where VMS must be used

Vessel Monitoring Systems (VMS) is a general term to describe systems that are used in commercial fishing to allow environmental and fisheries regulatory organizations to track and monitor the activities of fishing vessels. They are a key part of monitoring control and surveillance (MCS) programs at national and international levels. VMS may be used to monitor vessels in the territorial waters of a country or a subdivision of a country, or in the Exclusive Economic Zones (EEZ) that extend 200 nautical miles (370.4 km) from the coasts of many countries. VMS systems are used to improve the management and sustainability of the marine environment, through ensuring proper fishing practices and the prevention of illegal fishing, and thus protect and enhance the livelihoods of fishermen.

The exact functionality of a VMS system and the associated equipment varies with the requirements of the nation of the vessel's registry, and the regional or national water in which the vessel is operating. Within regional and national VMS initiatives there are also sub-divisions which apply different functionality to different vessel categories. Categories may be size or type of vessel or activity. For example:

- Local/regional fish such as scallops in the Northeast U.S., anchovies in Peruvian waters, or rock shrimp in the Gulf of Mexico

- Highly migratory species (HMS) such as tuna and billfish, or Patagonian toothfish (*Dissostichus eleginoides*) in the Antarctic. which can be caught in multiple regions

In this discussion, VMS relates specifically to fisheries management systems. VMS describes the specific application of monitoring commercial fishing boats. It is VTS which is describes the specific application of monitoring marine traffic primarily for safety and efficiency in ports and busy waterways. It is also not to be confused with specific communication technologies such as AIS, Iridium, Inmarsat, Argos, GPRS which relate to the communication method on which data is transmitted. Different VMS systems will use different communication technologies depending on the functionality requirements imposed by a national or regional VMS initiative.

The cost of VMS components will vary according to the functionality requirements of the specific system being implemented. In general the higher the functionality the more expensive the equipment and required data link (airtime costs). The cost of a VMS system therefore varies and thus the level of government subsidy (if any) varies according to national and regional requirements. EU and US VMS systems require expensive onboard equipment and large amounts of data to be transmitted over satellite link resulting in high airtime charges, but also provide a very high level of functionality. In other regions where per vessel cost and huge fleet sizes are an issue, communication technologies such as AIS are used which significantly reduce equipment and airtime costs whilst delivering acceptable basic VMS system functionality.

Applications

VMS is intended principally for fisheries management, but the country using it may use the data for other purposes.

Fisheries Management

Among all the most basic purposes is to monitor the movement of VMS-equipped vessels with respect to restricted fishing areas. A given vessel may have approval to fish in a restricted area, to transit through it without fishing, or it may not be allowed in the area.

Catch Reporting

Not all VMS systems are required to record and transmit catch reports. On the systems with separate PCs, it is reasonably easy to distribute separate software, although the fishermen may or may not be able to install it without assistance and additional equipment. For VMS with dedicated PCs, additional software can be made part of software support with established channels.

In the EU and US there is a distinct trend to making catch reporting part of an overall MCS pro-gram, although current VMS systems rarely integrate this capability with position reporting. One system that does to this is Fulcrum Maritime Systems Limited's Vessel Tracking Service which uses the latest data rich satellite application technology to enable electronic catch reports to be sent from the vessel to the VMS directly and automatically sends the report and vessel position data to the appropriate RFO or Regulatory Fishing Body, such as the NEAFC.

Under the European Union scheme, vessels are generally required to report

- Catch on entry

- Weekly catch

- Transshipment

- Port of landing

- Catch on exit

Within the EU an Electronic Reporting System is being implemented as part of the EU VMS system to automate collection of catch data, and exchange of data between EU states. A number of programs require tracking of days at sea (DAS) for a given vessel. They may require tracking the total cumulative catch of a given fishery.

Fisheries Research and Analysis

VMS is a planning and analysis tool as well as an aid to operations. Treated as a research database, the cumulative position reports gives an analysis of fishing vessel tracks in search of fish. Coupled with species-specific fishing licenses and catch reports, fisheries managers can estimate the amount of a given fish in an area, the amount taken by fishing vessels, and project overfishing before it happens.

Safety

VMS itself can help in search and rescue (SAR), especially when the SAR organization participates in the Global Maritime Distress Safety System(GMDSS). Some VMS have built-in Emergency Position-Indicating Radio Beacons (EPIRB), or SART capability although a dedicated VMS unit may not be able to have an emergency beacon that automatically floats to the surface and starts transmitting when it detects it is in salt water. At the very least, the SAR agency can get a last reported location of the vessel, and perhaps its course, from the FMC.

Enforcement

VMS obviously is part of fisheries enforcement, but, along with other systems, it can be part of overall sea surveillance. When a radar or other sensor detects a given vessel, VMS can tell the center that monitors the radar whether the radar target is a known fishing vessel. There may be correlation between AIS/VTS and VMS.

Technologies and Components

Any vessel tracking system, including VMS, requires technology on the vessel, ashore, and communications between them. In addition, there may be additional communications from the Fisheries Management Center (FMC) of the vessel's country of registry, and regional or national FMCs of the waters in which the vessel is fishing.

Functions Aboard The Vessel

The most basic function of a VMS is to determine the vessel's location at a given time, and periodically transmit this information, to a monitoring station ashore. Different VMS systems use different communication technologies, including AIS, Inmarsat, Iridium and Argos depending on the functionality required by the particular VMS system.

VMS components on the vessel sometimes are called VMS, or sometimes Automatic Location Communicators (ALC). These minimally include a GPS antenna and receiver, a computer (which may be embedded or user-supplied), and a transmitter and antenna appropriate for the communications that links the vessel to the flag center.

In practice, many of the VMS components also have applicability, along with non-VMS marine electronics, to a wide range of functions aboard a fishing vessel. These include navigation, finding fish, collision avoidance, routine voice and email communications, etc.

Selecting a VMS system is most dependent on what vendors and models have been approved by the fishing vessel's state of registry and the functionality requirements. Normally an authority will specify specific approved equipment to ensure end to end system integrity and service level meeting the specific requirements applicable to the vessel type. For example, some systems require a user interface on the vessel, whilst others will have a simple black box transceiver with no user interface.

VMS software and devices for the fishing vessel include:

- Absolute Software

- AMS

- Argonet-vms CLS

- AST Ltd

- Globavista/Bluefinger/Thales

- BlueTraker

- Boatracs

- CLS America - Thorium VMS

- Faria

- Free Port - Eye from a sky

- Fulcrum Maritime Systems - Vessel Tracking, Fleet Tracking, Fisheries Monitoring

- Honeywell

- Skymate

- SRT Marine System Solutions

- Thrane & Thrane

- vTrack VMS

- JouBeh Technologies iTrac VMS

- GMV

Communications

VMS units principally rely on global navigation satellite systems (GNSS) such as GPS for position and time information. LORAN may be a backup or complementary technology. These transceivers transmit data to monitoring systems generally using a variety of communication technologies including terrestrial & satellite AIS and conventional satellite systems from Inmarsat, Iridium, Argos, ORBCOMM or Qualcomm. Increasingly nations are implementing a mixture of technologies with the largest vessels over 60 tons being required to use the expensive traditional satellites and smaller vessels to use AIS.

- Operated by Inmarsat plc, originally founded by governments but now commercial, Inmarsat has a constellation of geosynchronous communications satellites.

- Iridium uses a constellation of 66 Low Earth Orbit satellites to provide complete global coverage (including all ocean regions and both poles) with real time coverage.

- Automatic Identification System - AIS is an IMO supported technology- provides low cost dual terrestrial and long range satellite maritime data communications. It offers a base level of VMS functionality with global coverage at the lowest cost.

- Argos uses Low Earth Orbit European and US satellites in polar orbit, which is an especially appropriate orbit for vessels operating in high latitudes.

- Skymate uses Orbcomm LEO satellites, which is optimized for machine-to-machine communications, potentially at lower cost than voice-capable satellite systems. They operate in the VHF and UHF bands, and have demonstrated an AIS capability.

- Qualcomm provides access to the Iridium satellite systems.

- BlueTraker uses both GPRS and Iridium constellation to provide the biggest flexibility and the lowest communication costs. The BlueTraker is a stand-alone device fully integrated including the antennas, the communication modules and a back-up battery. It is also e-logbook ready.

- iTrac is certified by the Canadian Department of Fisheries and Oceans, has integrated GPS and uses the Iridium satellite system network to provide communication as well as e-mail access.

VMS/ALC Vendor	Product type	Communications	Communications type
Applied Satellite Technology	Integrated Hybrid device with internal back-up battery, antennas and e-logbook capabilities	Dual mode GPRS & Iridium communications	VMS+ device Polar LEO with satellite-to-satellite
Blue Oceans Satellite Systems	GPS; duplex communication; Web-based monitoring interface	Iridium and Globalstar	Low-earth orbiting (LEO), cross-linked satellites operating as a fully meshed network and supported by multiple in-orbit spares.
Boatracs	Dedicated with phone, fax, email	Qualcomm	LEO L-band uplink
CLS	Argos	GPS with uplink	Polar LEO with satellite-to-satellite
Fulcrum Maritime Systems	Web based tracking for maritime or land based assets, Fisheries Monitoring Centre,	Open - Inmarsat-C, Skywave, Satamatics, Iridium, Globalstar	Systems Integrator, Bespoke system design, Supplier of tracking systems and devices, Simplex or Duplex, Ring Fencing, Movement detection, Global Coverage.
Globavista	Web based tracking, Ship Security Alert System & telemetry system	Open - AIS, Inmarsat-C, Skywave, Iridium, GSM/GPRS, Hybrid devices	Integrator/Supplier of majority of tracking devices
Satrax ETS-1000	Dedicated computer with telephone	Iridium	Polar LEO with satellite-to-satellite
Satrax ETS-250	Dedicated computer	Orbcomm	Polar LEO with satellite-to-satellite
Skymate	GPS & uplink antenna; PC software for VMS, weather, fish prices, surface temperature, log	Orbcomm	LEO VHF Uplink
SRT Marine System Solutions	Optimal AIS based VMS system for fishing vessels below 60 tons. High functionality combined with low cost and sophisticated display and data management systems which fuse terrestrial and satellite AIS to provide full EEZ and if required, global, coverage.		

exactEarth	A operator of a global network of low orbit AIS satellites which use technology such as ABSEA to receive transmissions from Class B and Identifier type AIS VMS transceivers.		
Thrane & Thrane	Dedicated computer with voice and email	Inmarsat-C	Geosynchronous satellite
Honeywell or Skywave	Low-cost, small transceivers with integrated GPS	Inmarsat-D+	Geosynchronous satellite
JouBeh Technologies Inc.	Low-cost, DFO Canada Certified, also gives users the ability to communicate via email over Iridium	Iridium	LEO Satellite Communications
EMA - BlueTraker	all-integrated Hybrid device (GPRS and satellite communications)with internal back-up battery, antennas and e-logbook capabilities	GPRS & Iridium	Polar LEO with satellite-to-satellite
Free Port - Eye from a sky (Maestral 2009)	integrated black-box device (GPS positioning, GPRS/EDGE and satellite communications) with internal back-up battery, external antenna and PC software for monitoring, surveillance, communication and data gathering	GPRS/EDGE & Iridium	Polar LEO with satellite-to-satellite

Fisheries Management Center of the Vessel's Nation

Software at the fisheries management organization looks for several pieces of information:

- location vis-a-vis restricted area

- time at sea

- time in restricted area, possibly separating fishing and transit time by speed

A restricted area may be closed for all purposes, open to transit, open to fishing for a specific period, or open to designated vessels only. Vessel speed is often the way its status is determined in lieu of direct observation. Some VMS directly report speed, or speed can be calculated by FMS software based on the time stamps of different position reports. A rule of thumb in scallop fisheries, for example, is that the vessel cannot be dredging for scallops if its speed is greater than 5 knots (9 km/h).

FMC software can note the time a vessel leaves and returns to port, and the time it is inside or outside designated areas. There may be restrictions on trip length, time in an area, etc., which can be calculated directly from VMS data. Other observations may require correlation of catch reports with the vessel's presence in given areas. Presence in other areas may require an onboard observer.

Individual regional, national, and international FMCs have different levels of software intelligence, which can detect patterns of interest to SAR, fisheries management, or law enforcement.

Fisheries Monitoring Center of the Waters Being Fished

Countries with registered fishing vessels that employ VMS generally agree to set up a Fisheries Monitoring Center (FMC), which has a data network connection to the FMCs of other states as well as other maritime stakeholders within the country. This flag state principle requires all vessels, registered in a given state to transmit their positions automatically to that state's FMC. When the vessel enters the waters of a different state, the home FMC must forward the report of the vessel's entry into those waters to the foreign state FMC. Until the vessel leaves the foreign state's coastal area, the home FMC must forward to the foreign state FMC the position, speed, and course reports at least every two hours. Exchange of data between VMS servers within the EU is regulated by the European Commission to be formatted according to the NAF FORMAT, originally devised by NAFO and subsequently adopted in a slightly different format by NEAFC. VMS servers outside of the EU may optionally also use the NAF format due to its widespread use within the EU. Data is usually transmitted using HTTPS protocol either by an HTTP Post or an HTTP Get request. Other protocols such as X.25 have historically been used but are in decline.

The NAF format is being replaced by FLUX, which defines messages and forwarding rules for many messages related to various business processes associated with VMS operation.

If position reports unexpectedly cease from a vessel, the FMC for the ocean area from which the last signal was received must attempt to contact the vessel or the flag state FMC without delay. Since VMS reports are sent automatically, it is possible that there is nothing wrong with the vessel itself, only the VMS. A full search and rescue (SAR) operation should not be launched simply because a VMS report does not arrive, although it is reasonable to alert sea surveillance assets, such as radars, that might be able to find the vessel. Fishing vessel crews should check the VMS at reasonable intervals, and confirm it is working.

While the procedure will vary with the jurisdiction, if an at-sea vessel finds their VMS is not working and they cannot fix it, they may be able to contact the FMC and get permission to continue the voyage. If they do get such authorization, they may get an inspection when they return to port. The FMC may also order them back to port. It is unlikely they will be allowed to leave port again without the VMS being repaired, so that they may need 24/7 VMS technical services at their home port.

Position reports received by the FMC should be forwarded automatically to the FMC of the vessel's registry. FMCs and other organizations, such as SAR and research, which receive VMS data must comply with confidentiality agreements. All recipients of data are also in accordance with agreements obliged to handle the data they receive in a responsible manner.

Catch Documentation Scheme

Catch reports are not themselves part of VMS, but often will be correlated with VMS data as part of an overall fisheries MCS program.

VTS and GMDSS

These are specific safety and traffic management systems which generally do not include specific fishery functionality as required by VMS.

AIS

AIS is often confused as a system comparable to VMS or VTS. AIS is a communications technology which is normally used within VTS and VMS applications. AIS is typically used on VMS systems deployed on smaller fishing vessels under 60tons. The implementation of AIS as part of these systems is also often customized to include encryption and other functionality.

International programs

Given that fish exist in food chains, it is worth nothing that the United Nations is at the logical top of the VMS chain, under the authority of the United Nations Convention on Law of the Sea (UN-CLOS). While it does not contain any provisions that are directly related to the use of VMS, it establishes a number of important principles of relevance for this study, relating to the conservation and management of living resources, both within national jurisdictions and on the high seas. UN fisheries operations are under the Food and Agricultural Organization.

Antarctic

Under the Commission for the Conservation of Antarctic Marine Living Resources (CCAMLR), a number of member states monitor agreed-to conservation measures and research information. The major emphasis is on the *Dissostichus* sp. catch (i.e., Patagonian toothfish and Antarctic cod) also known as Chilean sea bass. Realtime VMS is required for most exploratory VMS, with delayed reporting for other longline fisheries and for finfish trawling. The flag state VMS architecture is used, where the national FMCs of vessel registry, starting in 2005, transmit to the CCAMLR regional FMC.

Member states under the agreement are Argentina, Australia, Belgium, Brazil, Chile, European Community, France, Germany, India, Italy, Japan, Republic of Korea, Namibia, New Zealand, Norway, Poland, Russian Federation, South Africa, Spain, Sweden, Ukraine, United Kingdom of Great Britain and Northern Ireland, United States of America, and Uruguay. In addition, Bulgaria, Canada, Finland, Greece, Netherlands, Peru,and Vanuatu accede to the convention. In practice, up to 50 vessels are expected to be monitored, and about 15 of the convention states actually fish in the area. The longline vessels range from 349 to 2,203 long tons (355 to 2,238 t).

Depending on the latitude, satellite communications may require polar-orbiting satellites, or the vessel may be in line of sight of a geosynchronous satellite.

Europe

Under the European Union legislation, VMS is a legal requirement for vessels in excess of 15 metres. By 1999, Europe had 7000 vessels, in excess of 15 meters, under VMS. Since 2005, all Community vessels automatically transmit vessel identification, date, time, position, course and speed either hourly or every 2 hours (if the responsible Fisheries Monitoring Centre can request positions). The only exception is for vessels that operate only inside home waters, and are used exclusively for aquaculture.

One of the challenges for European MCS is that the idea of a 200-nautical-mile (370 km) EEZ is meaningless for nations with coasts in small seas such as the Mediterranean or Baltic. In such circumstances, appropriate international agreements need to be developed to govern fishing beyond the territorial limit and thus on the high seas, but high seas that would have overlapping jurisdiction in an EEZ-based model.

There are precedents where maritime pollution already is handled on a basin basis, which might provide a framework for fisheries enforcement in international waters of a small sea:

- Mediterranean (Barcelona Convention)

- Baltic (Helsinki Convention)

- North Sea (Bonn Agreement)

A recent IMO regulation requires AIS transponders aboard vessels, which transmit identifier, name and position of vessels not limited to fishing. Another approach might involve either AIS, or the more finely grained VTS, agreements that use coastal radar to monitor ships in and beyond coastal waters. This allows a transport vessel, for example, to be tracked in the small sea.

Another cross-check could involve current EU RADARSAT ScanSAR imagery, correlated with transponder information, such that only radar images of ships without active transponders display to enforcement personnel. At present, however, inspectors on aircraft or surface patrol vessels may not have real-time access to satellite imagery. Currently, the fusion of VMS, radar (satellite, aircraft, or coastal) has to be done at an operations center ashore. Another complication is that enforcement organizations for such things as spill monitoring are not concerned with issues such as illegal fishing.

In order to coordinate the policy making and enforcement efforts the CFCA - Community Fisheries Control Agency in Vigo, Spain has been established. The operational cooperation between Member States is organized with Joint Deployment Plans (JDPs). In order to support the JDPs, CFCA is operating a Vessel Monitoring system that in its first two years of operation (2009–2011) has exchanged 8 million VMS messages from 4520 vessels of 49 Flag States. The VMS software used by CFCA is vTrack.

Northwest Atlantic

The Northwest Atlantic Fisheries Organization is composed of Canada, Cuba, Denmark (in respect of the Faroe Islands and Greenland), European Union, France (in respect of St Pierre et Miquelon), Iceland, Japan, Republic of Korea, Norway, Russian Federation, Ukraine, United States of America.

Under this agreement are all fisheries, principally trawl and longline, except crab, lobster, salmon, sedentary species, whale and tuna. Approximately 135 vessels are monitored, the majority of which are trawlers with a few longline, ranging from 500 to over 2000 gross weight tons (GWT).

The VMS software used is vTrack.

Northeast Atlantic

The North-East Atlantic Fisheries Commission (NEAFC) has five Contracting Parties, Denmark (in respect of the Faroe Islands and Greenland), the European Union, Iceland, Norway and the Russian Federation.

This Convention (1980) regulates fisheries in the high seas (Regulatory Area) and waters under national jurisdiction, including trawlers, purse seiners, longliners, and gill netters. VMS is required, since 1 July 1999, for any vessel of 24 meters or longer overall length. NEAFC is a secondary user of data receiving it from flag state FMCs; the NEAFC database connects to national FMCs of Germany, Denmark, Spain, Estonia, France, Faroe Islands, United Kingdom, Greenland, Ireland, Iceland, Latvia, Lithuania, Netherlands, Norway, Poland, Portugal, Russian Federation, Sweden and the European Fisheries Control Agency. It is also connected to Co-operating Non-Contracting Parties FMCs of the Bahamas, Liberia and St Kitts and Nevis.

In 2004, 1473 vessels were monitored, with 800 authorized to fish for regulated species (Regulated Resources).

Fisheries surveillance platforms (vessels and aircraft) will also transmit information on their operations in the Regulatory Area, including:

- surveillance entry

- observations

- surveillance exit

NEAFC participated as an observer in, a project funded by the EU to use satellite radar images to validate VMS information and to complement and optimise surveillance tasks. NEAFC decided, based on the results of the EU-funded IMPAST (Improving Fisheries Monitoring By Integrating Passive and Active Satellite Technologies), to deploy a Vessel Detection System (VDS) in several coastal states. NEAFC also participated in the EU-funded SHEEL (Secure and Harmonized European Electronic Logbook) and CEDER (Catch, Effort and Discard Estimate in Real-time) projects, which may lead to direct electronic reporting of real-time catch data.

Pacific Islands

The FFA has 16 country members and one territory member from the western and central Pacific region. It is based in Honiara, Solomon Islands. While the FFA proper was formed over 20 years ago, VMS operation began in late 1997, covering the EEZs of 16 South Pacific countries.

FFA VMS is expected to cover 2000+ vessels, transmitting via Inmarsat-C and reporting every 4 hours.

Southern Africa

Fisheries are a major component of the economies of the coastal member states of the SADC (Angola, Namibia, South Africa, Mozambique, Tanzania, Mauritius and Seychelles).

Due to limited resources, there is little VMS beyond experiments in Namibia and South Africa. There is a European Union funded project to improve monitoring.

West Africa

The Sub-Regional Fisheries Commission (SRFC) based at Dakar, Senegal is made up of west African States, namely Cape Verde, The Gambia, Guinea, Guinea-Bissau, Mauritania, Sierra Leone and Senegal. Its role is to promote fisheries cooperation between its member States.

The donor is the Grand Duchy of Luxembourg and the executing agency is the FAO and Lux-Development. Participating countries are These countries are members of the Sub-Regional Fisheries Commission (CRSP) with the addition of Sierra Leone. The project proposed is a continuation and extension of the AFR/101 Project (FAO), which may add VMS to supplement the present air surveillance.

National

This section deals with the specifics of national use of VMS, rather than their overall approach to fisheries management.

Albania

Albania is currently implementing a VMS for its 220 vessels fishing fleet, combining satellite and GPRS communications channels. The BlueTraker VMS solution, which is E-Logbook ready, is supplied by company EMA..

Argentina

The Commission for the Conservation of Antarctic Marine Living Resources demands the consensus of its 24 member countries for any proposals to be implemented. At its annual meeting in Hobart over the past two weeks, Argentina could not be persuaded to approve the adoption of a centralised system to monitor pirate ships.

The Parliamentary Secretary to the Minister for Environment, Sharman Stone, says Argentina was suspicious of the technology. "Argentina was concerned that we couldn't guarantee the confidentiality of any system, now the technical requirements of such a system were agreed that this wasn't beyond anyone's technical ability and capacity, but unfortunately Argentina remained concerned about the confidentiality of the data," she said.

However, the commission has agreed that Australia and the United States will head up a trial of the centralised vessel monitoring system over the coming season.

Australia

Australia has both national and state programs. The national-level program is run by the Australian Government agency, the Australian Fisheries Management Authority (AFMA). VMS runs on about 500 (growth expected to 800) vessels from small 10-meter scallop boats to 850-meter deep sea trawlers.

Fisheries of interest include orange roughy, scallops, prawns, tuna and billfish. Fishers must use AFMA-approved VMS devices.

Southern Australia

There is a regional organization of Southern Australian states which monitors rock lobster, giant crab, and, on a sampling basis, aquaculture.

Canada

Since 2001, Canada has mandated VMS, for vessels of certain sizes, to fish for specific species in designated areas. The underlying MCS strategies, while differing in specific fisheries, are based on limited entry licensing, with restrictions on vessel and gear types. Canada expects VMS reports every two hours.

Canadian activities involving VMS are joint between the Department of Fisheries & Oceans (DFO) and the Department of National Defense (DND). DND is the lead department for an inter-Departmental web-based mapping application, supported by positional information from DFO.

DND provides non-VMS surveillance data to a DND-operated data base available to DFO for fisheries management. Aerial surveillance, using a variety of sensors, monitors freighters, tankers, bulk carriers and container ships as well as fishing vessels.

Canada intends to provide, in real time, VMS data, correlated with other sensors and database information, to patrol vessels and aircraft. Electronic logs, two-way communication with fishing vessels, issuing orders, and possibly placing video and other sensors on fishing.

Chile

Chile has VMS aboard approximately 500 vessels, each vessel being associated with one fishery.

Marimsys built the Chilean VMS, but that system was replaced in 2007 with a new one provided by CLS.

Chile went from a government-specified VMS to a type-approval system for VMS units in an open market, which drove down costs.

Monitored industrial fishing boats limit fishing activities to—generally—5 nautical miles (9 km) from the coast of Chile. This leaves the 5 nm zone for "artisanal" or smaller fishing boats and limits excessive fishing effort being applied to inshore waters.

Chile also pioneered in the emission of reports at short intervals. Prior systems had focussed on "where is the vessel" with the provision of hourly reporting. The Chilean system, by dropping the minimum report interval to 8 minutes is capable of determining "what" the vessel is doing. When you see a series of circular positions, they are all at speeds of below 2 knots (4 km/h) and reflect the drift of the current—there is no question, that vessel was purse seining, and the printout of the chart can be shown to the court to demonstrate the fact.

Chile is currently the world's largest producer of farmed salmon and has a burgeoning mussel culture industry that is supplying a growing world market. Other fisheries of interest include alfonsino, anchovies, cod, cuttlefish, hake, mackerel, ray, sardines, sea bream, squid, and swordfish.

The system is also used to monitor foreign vessels entering and leaving both the EEZ and Chilean ports.

Taiwan

Taiwan has national and provincial VMS programs, the most active being for Taiwan. It uses both Inmarsat-C and Argos to monitor up to 1200 vessels.

Croatia

Croatia has implemented its VMS on 256 vessels in 2007. The BlueTraker VMS devices, supplied by EMA, enable utilization of both satellite and GPRS communication channels. VMS software solution is developed and supported by GDI GISDATA, a Croatian company. Croatian Fisheries department uses it to identify and track the country's large fishing vessels. This information can be used for monitoring boat activity and as evidence for law enforcement.

The main components of VMS are the department's centralized database, tracking devices, and ArcGIS. Whether at the department, in the harbor office, or on a boat, an inspector can access the GIS to track a vessel and get information about its owner, type, and gear on board and a host of other information. VMS collects vessel information in real time, such as location, speed, direction, and even battery status. Developed on ArcGIS for Server using the ArcGIS API for JavaScript, the system integrates with vessel data stored in the Microsoft SQL Server database and publishes dynamic content.

Denmark

Denmark has a nationwide VMS based on Inmarsat-C transceivers owned and maintained by the authorities. The VMS software is vTrack. The system monitors 600 vessels.

Ecuador

Ecuador uses VMS for tuna, under the Association of Tuna Fishing Companies of Ecuador (ATUNEC).

Estonia

Estonia has a nationwide VMS based on Inmarsat-C hardware and vTrack software. The system monitors 50 vessels. The Estonian VMS system is operated by the Estonian Environmental Inspectorate. The VMS software is vTrack.

Falkland Islands

The Falkland Islands has a VMS program for all vessels licensed to fish in its waters.

Faeroe Islands

Faeroe Islands has a nationwide VMS based on Inmarsat-C hardware. The system monitors 150 vessels. The VMS software is vTrack.

France

Implementing its FMC at the CROSS sea rescue center at Etel, France uses the flag state principle described under Norway. The MAR-GE unit is a self-contained GPS and Argos device. France expects 2-hour reporting.

Germany

The German VMS is based on Inmarsat-C transceivers. The VMS software is vTrack. The system monitors 300 vessels.

Greenland

Greenland VMS is based on Argos/CLS and Inmarsat-C transceivers. The VMS software is vTrack. The system monitors 100 vessels.

Iceland

Iceland uses VMS for both safety and fisheries compliance, monitoring with Inmarsat-C or a coastal VHF repeater system. Approximately 1600 vessels of all sizes are monitored. Thales VMS has been approved.

India

India is introducing VMS for its EEZ, along with a system of permits to control capacity.

Indonesia

The Indonesian Ministry of Maritime Affairs and Fisheries selected Argos for their VMS. Indonesia's VMS system is the largest, or among the largest, in the world. 1500 fishing vessels initially with VMS, with three ashore FMCs. A distinctive feature of the Indonesian system is that an initial 15 patrol boats can directly receive VMS information.

Ireland

The Irish VMS system is operated by the Irish Naval Service, based in County Cork. As well as monitoring Irish vessels, the VMS exchanges data with VMS systems operated by other EU states.

Japan

A framework for groundfish fisheries in the Northwest Pacific's high seas was established in January 2007 in a consultation involving Japan, Russia, South Korea and the United States. VMS will be used to collect data.

Lithuania

Lithuania has a nationwide VMS based on Inmarsat-C hardware. The system monitors 50 vessels. The VMS software is vTrack.

Malaysia

Malaysia uses VMS on its Malaysian Maritime Enforcement Agency patrol boats and also on larger fishing vessels, through the Fisheries Department.

Malta

Malta monitors approximately 60 vessels.

Mexico

Under current Mexican law, it is illegal for commercial boats like longliners and drift gillnetters,to take fish reserved for sports fishing within fifty miles (93 km) of the coast in the Sea of Cortez, and any fish within 12 nautical miles (22 km) of the Revillagigedo Islands. VMS is seen as the only way Mexico will to enforce controls on areas in its EEZ.

Morocco

Morocco are currently implementing a VMS system combining satellite tracking and radar correlation, supplied primarily by BlueFinger Ltd.

Namibia

The fisheries in Namibia are among the largest in Africa, with some of the most sophisticated MCS systems.

VMS is fully operational and has been implemented across many fishing fleets. Following an EU funded MCS program for the SADC region, Namibia has facilities to integrate its VMS data with that of other SADC partners so that information can be shared regarding vessels that operate across the border in another SADC states waters. Similarly, Namibia can receive VMS information from its SADC partners when a vessel from another SADC state enters its waters.

The observer program has been effective. Nevertheless, it may be appropriate, initially for the orange roughy fishery.

Nauru

All foreign vessels licensed to fish or support fishing operations in Nauru waters are required to use an Automatic Location Communicator compatible and registered with the Regional Vessel Monitoring Systems serving the Pacific Islands Forum Fisheries Agency and the Western and Central Pacific Fisheries Commission. Nauru has VMS data-sharing agreements with several other FFA member countries. A list of vessels licensed to fish in Nauru fisheries waters is uploaded daily to the FFA website.

Netherlands

The Netherlands has a nationwide VMS based on Inmarsat-C hardware. The system monitors 500 vessels. The VMS software is vTrack.

New Zealand

New Zealand has been running VMS since April 1994, with coverage out to the EEZ border under national and state agencies, with a target of 1000 vessels reporting every 2 hours. National & State Fisheries Agencies are responsible for the management of Fisheries located within its EEZ. Vessels use either Argos or Inmarsat-C to report position every 2 hours

Norway

Norway requires VMS aboard all of its fishing vessels longer than 15 meters. Norway has established such a centre at the Directorate of Fisheries in Bergen. Norway currently has mutual tracking agreements with the EU, Russia, Iceland, the Faeroe Islands and Greenland.

Panama

When Panama joined the International Commission for the Conservation of Atlantic Tunas (IC-CAT) in 1998, in response to an ICCAT embargo on bluefin tuna, it committed to require licensing and equipping deep sea fishing vessels aith the Argo ELSA VMS.

Peru

Peru uses VMS to manage its anchovy fishery. For Peru fishing is a prime source of foreign exchange, second only to mining. Over 1000 fishing vessels are tracked in Peruvian waters by Argos. The Peruvian government implemented a national fishing Vessel Monitoring System (VMS) in 1998 to monitor and track all fishing vessels in its Exclusive Economic Zone. One of the first major VMS system's in the world, the system continues to operate today and is a reference for other countries wishing to implement similar fisheries management capabilities.

The country's anchovy fishing fleet, which seeks the Peruvian anchovy *Engraulis ringens*, is the world's largest single-species fishery, with an average of 8% of global landings.

For research, safety and monitoring purposes, vessels have the statutory obligation to use VMS, with industrial-scale fishing prohibited within 5 nautical miles (9 km) from the coast.

Poland

Poland has a nationwide VMS based on Inmarsat-C hardware. The VMS software is vTrack.

Russia

The Russian Federation has an integrated system called SSM, for fisheries resource monitoring and has implemented a sectoral system for monitoring of the aquatic living resources, and for surveillance and control over the activities of the fishing vessels (SSM). SSM includes VMS monitoring of vessel positions.

SSM headquarters is in Moscow, with regional monitoring centers in Murmansk and Petropavlovsk-Kamchatskiy. The national system covers approximately 3800 vessels. Bilateral agreements exist with Faroes, Greenland Iceland, Japan, and Norway. Russia participates in the NAFO, NEAFC, and CCALMR multinational agreements. It regards SSM as integral to safety of navigation and SOLAS. Russia has bilateral agreements with Japan. AMS builds a Russian VMS.

Kamchatka Region

This covers the Pacific Ocean and the eastern Arctic Sector.

Murmansk Region

The Murmansk region covers Russian vessels in the Atlantic Ocean, the Azov, Black and Caspian Sea regions, and the western Arctic Sector/

Slovenia

Slovenia has a nationwide VMS based on Inmarsat-C hardware. The system monitors 8 vessels. The VMS software is vTrack.

South Africa

Fisheries management, including limited VMS, is under the Marine and Coastal Management (MCM) organization in the Department of Environmental Affairs and Tourism. Hake *Merluccius spp.*) trawl fishery is the mainstay of South Africa's fishing industry, and the center of regulatory efforts. On-board observers had been the mainstay of monitoring, rather than VMS.

VMS is aboard many vessels with reporting to an FMC in Cape Town that is equipped with Blue-Finger's VMS software. Additional VMS will go onto vessels into more distant waters, such as hake longliners. VMS is seen as a management, a research, and a safety tool. South Africa is exploring correlating its VMS with:

- RadarSat off Prince Edward Island, possibly in lieu of patrol vessels there,

- Airborne Synthetic Aperture Radar for quick-look surveillance and coverage out to the edge of the EEZ.

- Coastrad, a system of linked coastal radars for monitoring specific vessels, as verifying that foreign fishing vessels conducting innocent passage do that, rather than fish

- Patrol vessels to back up all other sensors.

South Korea

The Korean Squid Fishing Association has 102 fishing veesels which are to be equipped with an Argos satellite transmitter for their fishing campaigns in Russia.

Suriname

CLS/Argos is under the Ministry of Agriculture, Farming and Fisheries, and was initially deployed on over 80 vessels.

Sweden

Sweden has a nationwide VMS established in 1998 and based on Inmarsat-C. The system is operated 24/7 by Havs- och vattenmyndigheten (The Swedish Agency for Marine and Water Management) and monitors all fishing vessels >= 12 m LoA. Number of vessels being monitored: 184 [2013-05-22]. These vessels use electronic fishing logbook and the reports are automatically transmitted via mobile internet close to shore and via the VMS-equipment further out at sea. No exemptions from VMS and electronic fishing logbook are allowed for the time being.

United Kingdom

In the UK recently new requirements have been introduced through both EU and National Legislation requiring the use of VMS to monitor fishing fleets for both fishing effort as well as address the protection of Marine habitats. Parallel to this the EU has also introduced the use of electronic logbooks, which replaces the traditional use of paper records. The UK fishing authorities are made up of the Welsh Government, Department of Agriculture & Rural Development Northern Ireland, Isle of Man Department for Environment Fisheries & Agriculture, Marine Scotland, Marine Management Organisation and the Channel Island Authorities.

EU and National reporting schemes are defined thus: National VMS reporting schemes: These are fishing vessel position reporting requirements that form part of management schemes set up by one or more UKFAs to control certain fisheries and marine conservation areas. Commonly referred to as 'National VMS reporting', it uses both satellite and GPRS/GSM communication services depending on the Legislative requirements. Two schemes are currently operational: for Northern Ireland Mussel Dredging and Isle of Man Scallop Fisheries. EU VMS reporting scheme: these are fishing vessel position reporting requirements, through satellite communication services only, that enables the authorities to track the position of fishing vessels, as set out in EU Regulations. It is commonly referred to as 'EU VMS reporting'. In addition, electronic logbook requirements are defined thus: catch and effort recording in 'real time' but as a minimum every 24hrs whilst a fishing vessel is at sea. Catch and effort data can be transmitted over whichever communication channel is available (i.e., if the Satellite Communication Service is unavailable, connection will be attempted via GPRS/GSM and vice versa). Electronic logbooks are commonly referred to as 'E-logbooks' and the end to end process and accompanying software is referred to as 'Elogbook Software System or ELSS'. In order to address this challenge the UK fishing authorities teamed up with Applied Satellite Technology Ltd (AST Ltd) in 2012 to come up with a combined solution. The resulting solution was the VMS Plus device, which has the capability to deliver all of these requirements through one device and is now being rolled out to the UK fishing fleet. In summary the VMS Plus device meets the following functional requirements in full:

- Position reporting in accordance with the EU VMS regulatory requirements;

- Position reporting in accordance with relevant National Regulations governing marine protected and other special conservation areas at sea;

- Polling to request current and/or past positions from the device. The VMS Plus device has its own internal GPS used for position reporting and therefore is a useful cross referencing tool enabling fishing authorities to cross reference VMS positional data with other sources of positional data such as electronic logbooks;

- Separate access to communication services in the device, for other on-board systems, which as a minimum allow an on-board system using any UK fishing authority approved E-logbook Software to transmit E-logbook reports to, and receive acknowledgements from the UK fishing authorities.

Tunisia

Tunisian company "GEOMATIX" have developed a Vessels Monitoring System based on satellite communication and GPS position measurement. Tunisian's VMS solution provide pertinent information on the fishing fleet activities that help the administration manage the fishing vessels and control the sea resource and fishing productions.

United States

In the U.S., national fisheries management is under the National Oceanographic and Atmospheric Administration's (NOAA) Fisheries Service. There may also be state fisheries regulators. USA has a nationwide VMS monitoring a total of approximately 4.500 vessels based on the vTrack software.

Nationally-defined Fisheries VMS

At the national level, the goals include:

- Days at sea
- Electronic logbooks (at-sea catch reporting of species of interest)
- Area closures based on total allowable catch
- Spatial analysis of catch
- Measure compliance with restricted fishing areas, both domestic and international
- Area-specific quota management
- Measure compliance with EEZ boundaries by foreign vessels operating under settlement conditions

Northeast Region

This region consists of the Northeast Region includes marine waters off U.S. states of Maine, New Hampshire, Massachusetts, Rhode Island, Connecticut, New York, New Jersey, Pennsylvania, Delaware, Maryland, Virginia, North Carolina. Approximately 600 vessels are under VMS, with growth expected to 2500. VMS, operational since 1998, has been a significant tool in detecting trespass into closed areas by scallop and multispecies vessels, Days-at-Sea (DAS) violations, and fish landings that exceed trip limits (particularly in the cod and general category scallop fisheries). Numerous catch seizures and closed area violation prosecutions have been made solely from VMS data.

VMS is a basic tool in calculating DAS for the multispecies or DAS fisheries. Whenever a VMS fails to transmit an hourly function, it will be charged a DAS, unless the preponderance of evidence demonstrates that the failure to transmit was due to an unavoidable malfunction or disruption of the transmission, or was not at sea.

Fisheries of interest include:

- Scallop (dredge)

- Northeast multispecies (trawl, gillnet, hook gear)

- Monkfish (trawl, gillnet)

- Herring (trawl)

- Cod

Southeast Region

The Southeast Region includes marine waters off U.S. states of North Carolina, South Carolina, Georgia, Florida, Alabama, Mississippi, Louisiana, and Texas, as well as US waters around Puerto Rico and the US Virgin Islands.

Fisheries monitored by VMS

- Rock shrimp endorsement holders (trawl)

- Highly migratory species, or HMS (pelagic longline)

- Shark (gillnet and bottom longline gear)

- Penalty fleet (vessels subject to VMS monitoring as a penalty for violating fisheries regulations)

- Reef fish

269 vessels monitored by VMS

Skymate VMS is not approved for reef fish.

Northwest

The Northwest Region includes marine waters off U.S. states of Washington, Oregon and California.

- Rockfish Conservation Areas (RCAs): large-scale depth-based areas for stock preservation

- Cowcod Conservation Areas (CCAs): areas of previous overfishing

- Yelloweye Rockfish Conservation Area (YRCA): for another overfished species

The Pacific Coast vessel monitoring program consists of declaration reports and a vessel monitoring system. The declaration reports must be filed leaving port, and must identify:

- the vessel operator's intent to fish within an RCA,

- the gear type will be used for fishing,

- the fishery they are participating in.

Declaration reports are only necessary for fisheries that are allowed within a closed area and before a vessel intends to fish.

Southwest

Salmon-related issues remain the priority.

Alaska

Argos CLS is approved in this region.

Pacific Islands

Argos and Inmarsat-C are approved. The Pacific Islands Region includes waters around U.S. islands of Hawaii, Guam, Northern Mariana Islands, American Samoa, Wake Island, Midway Island, Howland and Baker Islands, Kingman Reef and Palmyra Atoll, Johnston Island and Jarvis Island. Approximately 200 vessels have VMS.

- Pelagic longline

- Northwestern Hawaiian Islands lobster trap

- American Samoa alia (small vessel longline, pilot project)

- Tuna purse seine (operating under South Pacific Tuna Treaty)

- Krill trawl (operating under CCAMLR)

- Foreign longline, pole and line (operating according to terms of court-ordered settlement agreements resulting from violations of US fishery law)

Highly Migratory Species in Multiple Regions

The VMS for HMS consists of both the mobile transceiver unit placed on the vessel and the communications service provider that supplies the wireless link between the unit on the vessel and the shoreside data user.

In the HMS Fisheries, the vessel owner is required to procure both VMS components. The two VMS components may, or may not, be provided by a single vendor. Thus, the vessel owner may need to procure the mobile transceiver unit and the mobile communications service separately.

VMS transmit vessel information primarily via INMARSAT satellites. They receive time and position data from the GPS constellation.

Uruguay

The authorities are theMaritime Authority, Dirección de Marina Mercante (DIMAR), and the Fishery Authority, Dirección Nacional de Recursos Pesqueros (DINARA). Uruguay licenses vessels for:

1. Hake *Merluccius hubbsi* on the continental shelf of the Uruguayan-Argentine Common Fishing Zone in depth more than 50 meters

2. white croaker *Micropogonias furnieri* and sea trout *Cynoscion guatucupa*, fishing in the coastal zone less than 50 meters depth in the La Plata River and the Uruguayan-Argentine common fishing zone

3. Various vessels different from 1 and 2, that can fish in Uruguayan waters.

4. Various vessels that fish in international waters. At present these vessels are fishing in FAO statistical area 41, CCAMLR 88.1 and 48.3, and the Pacific Ocean.

Commercial Fishing

Commercial crab fishing

Commercial fishing is the activity of catching fish and other seafood for commercial profit, mostly from wild fisheries. It provides a large quantity of food to many countries around the earth, but those who practice it as an industry must often pursue fish far into the ocean under adverse conditions. Large-scale commercial fishing is also known as industrial fishing. This profession has gained in popularity with the development of shows such as *Deadliest Catch*, *Swords*, and *Wicked Tuna*. The major fishing industries are not only owned by major corporations but by small families as well. The industry has had to adapt through the years in order to keep earning a profit. A study taken on some small family-owned commercial fishing companies showed that they adapted to continue to earn a living but not necessarily make a large profit. It is the adaptability of the fishermen and their methods that cause some concern for fishery managers and researchers; they say that for those reasons, the sustainability of the marine ecosystems could be in danger of being ruined.

Commercial fishermen harvest a wide variety of animals, ranging from tuna, cod, carp, and salmon to shrimp, krill, lobster, clams, squid, and crab, in various fisheries for these species. There are large and important fisheries worldwide for various species of fish, mollusks, crustaceans, and echinoderms. However, a very small number of species support the majority of the world's fisheries. Some of these species are herring, cod, anchovy, tuna, flounder, mullet, squid, shrimp, salmon, crab, lobster, oyster and scallops. All except these last four provided a worldwide catch of well over a million tonnes in 1999, with herring and sardines together providing a catch of over 22 million metric tons in 1999. Many other species are fished in smaller numbers.

The industry, in 2006, also managed to generate over 185 billion dollars in sales and also provide over two million jobs in the United States, according to an economic report released by NOAA's Fisheries Service. Commercial fishing may offer an abundance of jobs, but the pay varies from boat to boat, season to season. Crab fisherman Cade Smith was quoted in an article by *Business Week* as saying, "There was always a top boat where the crew members raked in $50,000 during the three-to five-day king crab season—or $100,000 for the longer snow crab season". That may be true, but there are also the boats who don't do well; Smith said later in the same article that his worst season left him with a loss of 500 dollars.

A 2009 paper in *Science* estimates, for the first time, the total world fish biomass as somewhere between 0.8 and 2.0 billion tonnes.

Methods and Gear

Seabirds with longline fishing vessel

Commercial fishing uses many different methods to effectively catch a large variety of species including the use of pole and line, trolling with multiple lines, trawling with large nets, and traps or pots. Sustainability of fisheries is improved by using specific equipment that eliminates or minimizes catching non-targeted species.

Fishing methods vary according to the region, the species being fished for, and the technology available to the fishermen. A commercial fishing enterprise may vary from one man with a small boat with hand-casting nets or a few pot traps, to a huge fleet of trawlers processing tons of fish every day.

Commercial fishing gears in use today include surrounding nets (e.g. purse seine), seine nets (e.g. beach seine), trawls (e.g. bottom trawl), dredges, hooks and lines (e.g. long line and handline), lift nets, gillnets, entangling nets, Pole and Line, and traps

Commercial fishing gear is specifically designed and updated to avoid catching certain species of animal that is unwanted or endangered. Billions of dollars are spent each year in researching/developing new techniques to reduce the injury and even death of unwanted marine animals caught by the fishermen. In fact, there was a study taken in 2000 on different deterrents and how effective they are at deterring the target species. The study showed that most auditory deterrents helped prevent whales from being caught while more physical barriers helped prevent birds from getting tangled within the net.

Occupational Risk

Trawl fishermen wearing personal flotation devices in a 2009 trial

During 2000–2006, commercial fishing was one of the most dangerous occupations in the United States, with an average annual fatality rate of 115 deaths per 100,000 fishermen. Falling overboard specifically killed 182 fishermen in the period between 2000 and 2010. This fatality rate is 3 times that of the next most dangerous job in the U.S. and more than 25 times that of the national average across all workers. Also, between the years of 1919 and 2005, 4111 fishermen died in fishing related accidents in the United Kingdom industry alone. These deaths are generally a result of a combination of severe weather conditions, extreme fatigue because any one fisherman usually puts in a 21-hour shift, and dangerous equipment. The U.S. Coast Guard has primary jurisdiction over the safety of the U.S. commercial fishing fleet, enforcing regulations of the U.S. Commercial Fishing Industry Vessel Safety Act of 1988 (CFIVSA). CFIVSA regulations focus primarily on saving lives after the loss of a vessel and not on preventing vessels from capsizing or sinking, falls overboard, or injuries on deck. CFIVSA regulations require that commercial fishing vessels carry various equipment (e.g., life rafts, radio beacons, and immersion suits) depending on the size of the vessel and the area in which it operates. Not all commercial fishermen follow safety regulations and advice.

One study of Maine fishermen found that less than 25% of the fishermen interviewed had recent training in first aid or CPR, only 75% of the boats had survival suits and only 36% had a survival craft. Even the ships that did have the necessary equipment did not consistently have a captain that fully understood how to use the safety equipment.

San Miguel Rescue - The Coast Guard Rescued Three Commercial Fishermen

Common causes of fishing-related deaths include vessel disasters, falls overboard, and onboard injuries. The United States National Institute of Occupational Safety and Health Commercial Fishing Incident Database found that between 2000 and 2010, most vessel disasters often were initiated by flooding, vessel instability, and large waves, and that severe weather conditions contributed to a majority of fatal vessel disasters. Most falls overboard went unwitnessed, and in none of the cases documented was the victim wearing a personal flotation device (PFD). Onboard injuries often result when a crew member is caught in a line and pulled into a winch on deck. The installation of a readily accessible emergency stop switch on the winch can potentially prevent these kinds of injuries.

Several institutions have tried to change the culture surrounding safety on commercial fishing boats, especially around wearing personal flotation devices. The Alaska Scallop Association mandates that every fisherman must wear a PFD while on deck of the boat, and other organizations have purchased more wearable PFDs.

Environmental Risk

The ocean covers nearly two thirds of the Earth's surface, and is continuously threatened by human behaviors and practices. By taking so many fish from the seas, humans have managed to remove entire links from the aquatic food chain. This causes a chain effect, leading to an overall upset of the delicate balance of nature.

Sharks are one of the ocean's most threatened species because they are mistakenly caught by vessels searching for fish, and end up getting tossed back into the ocean dead or dying This disappearance of sharks has enabled prey animals like rays to multiply, which alters the ecological food chain.

Bycatch is the industry term for what they consider "unwanted or economically-worthless aquatic animals who are unintentionally caught using destructively indiscriminate fishing methods like longlines and driftnets, which generally target marketable marine creatures such as tuna and swordfish" There are also billions of other animals that are killed in this manner every year such as: sea turtles, marine mammals, and sea birds. Between 1990 and 2008, it was estimated that 8.5 million sea turtles were fatally caught in nets or on longlines as bycatch.

Coral reefs have intense biodiversity-rich ecosystems which provides habitat for millions of aquatic species such as sponges, star fish, jellyfish, sea turtles, etc. Unfortunately, reef ecosystems are highly sensitive to chemical, temperature, and population changes. There has been an increasing disappearance of large predators such as barracuda, Nassau groupers, and sharks This makes the reefs more vulnerable to invasion by non-native species.

Fish farming is the raising of fish for food in underwater enclosures, otherwise known as aquaculture. There are environmental hazards such as waste, damage to ecosystems, and negative effects on humans. Because they are so densely packed together, the fecal matter that accumulates can create algal blooms, or deadly parasites and viruses that thrive on the filthy environment. These can infect wild fish that swim near the enclosure, or whole colonies of fish if an infected farm fish escapes the enclosure.

Overfishing occurs because fish are captured at a faster rate than they can reproduce. Both advanced fishing technologies and increased demand for fish have resulted in overfishing. The Food and Agricultural Organization has reported that "about 25 percent of the world's captured fish end up thrown overboard because they are caught unintentionally, are illegal market species, or are of inferior quality and size" It should not go unnoticed that overfishing has caused more ecological extinction than any other human influence on coastal ecosystems.

Artisanal Fishing

Stilts fishermen, Sri Lanka

Artisanal fishing (or traditional/subsistence fishing) are various small-scale, low-technology, low-capital, fishing practices undertaken by individual fishing households (as opposed to commercial companies). Many of these households are of coastal or island ethnic groups. These households make short (rarely overnight) fishing trips close to the shore. Their produce is usually not processed and is mainly for local consumption. Artisan fishing uses traditional fishing techniques such as rod and tackle, fishing arrows and harpoons, cast nets, and small (if any) traditional fishing boats.

Artisan fishing may be undertaken for both commercial and subsistence reasons. It contrasts with large-scale modern commercial fishing practices in that it is often less wasteful and less stressful on fish populations than modern industrial fishing.

Artisan Fishing Boats

Fishermen at work off the northern coast of Mozambique

Fishing boats at Koh Tao, Thailand

Artisan Techniques

Cormorants used for fishing in China

Fish Processing

Humans have been processing fish since neolithic times. This 16th-century fish stall shows many traditional fish products.

The term fish processing refers to the processes associated with fish and fish products between the time fish are caught or harvested, and the time the final product is delivered to the customer. Although the term refers specifically to fish, in practice it is extended to cover any aquatic organisms harvested for commercial purposes, whether caught in wild fisheries or harvested from aquaculture or fish farming.

Larger fish processing companies often operate their own fishing fleets or farming operations. The products of the fish industry are usually sold to grocery chains or to intermediaries. Fish are highly perishable. A central concern of fish processing is to prevent fish from deteriorating, and this remains an underlying concern during other processing operations.

Fish processing can be subdivided into fish handling, which is the preliminary processing of raw fish, and the manufacture of fish products. Another natural subdivision is into primary processing involved in the filleting and freezing of fresh fish for onward distribution to fresh fish retail and catering outlets, and the secondary processing that produces chilled, frozen and canned products for the retail and catering trades.

There is evidence humans have been processing fish since the early Holocene. These days, fish processing is undertaken by artisan fishermen, on board fishing or fish processing vessels, and at fish processing plants.

Overview

Fish is a highly perishable food which needs proper handling and preservation if it is to have a long shelf life and retain a desirable quality and nutritional value. The central concern of fish processing is to prevent fish from deteriorating. The most obvious method for preserving the quality of fish is to keep them alive until they are ready for cooking and eating. For thousands of years,

China achieved this through the aquaculture of carp. Other methods used to preserve fish and fish products include

- the control of temperature using ice, refrigeration or freezing

- the control of water activity by drying, salting, smoking or freeze-drying

- the physical control of microbial loads through microwave heating or ionizing irradiation

- the chemical control of microbial loads by adding acids

- oxygen deprivation, such as vacuum packing.

Usually more than one of these methods is used. When chilled or frozen fish or fish products are transported by road, rail, sea or air, the cold chain must be maintained. This requires insulated containers or transport vehicles and adequate refrigeration. Modern shipping containers can combine refrigeration with a controlled atmosphere.

Fish processing is also concerned with proper waste management and with adding value to fish products. There is an increasing demand for ready to eat fish products, or products that do not need much preparation.

Handling the Catch

Cleaning fish, 1887. By John George Brown.

When fish are captured or harvested for commercial purposes, they need some preprocessing so they can be delivered to the next part of the marketing chain in a fresh and undamaged condition. This means, for example, that fish caught by a fishing vessel need handling so they can be stored safely until the boat lands the fish on shore. Typical handling processes are

- transferring the catch from the fishing gear (such as a trawl, net or fishing line) to the fishing vessel

- holding the catch before further handling

- sorting and grading

- bleeding, gutting and washing

- chilling

- storing the chilled fish

- unloading, or landing the fish when the fishing vessel returns to port

The number and order in which these operations are undertaken varies with the fish species and the type of fishing gear used to catch it, as well as how large the fishing vessel is and how long it is at sea, and the nature of the market it is supplying. Catch processing operations can be manual or automated. The equipment and procedures in modern industrial fisheries are designed to reduce the rough handling of fish, heavy manual lifting and unsuitable working positions which might result in injuries.

Handling Live Fish

An alternative, and obvious way of keeping fish fresh is to keep them alive until they are delivered to the buyer or ready to be eaten. This is a common practice worldwide. Typically, the fish are placed in a container with clean water, and dead, damaged or sick fish are removed. The water temperature is then lowered and the fish are starved to reduce their metabolic rate. This decreases fouling of water with metabolic products (ammonia, nitrite and carbon dioxide) that become toxic and make it difficult for the fish to extract oxygen.

Fish can be kept alive in floating cages, wells and fish ponds. In aquaculture, holding basins are used where the water is continuously filtered and its temperature and oxygen level are controlled. In China, floating cages are constructed in rivers out of palm woven baskets, while in South America simple fish yards are built in the backwaters of rivers. Live fish can be transported by methods which range from simple artisanal methods where fish are placed in plastic bags with an oxygenated atmosphere, to sophisticated systems which use trucks that filter and recycle the water, and add oxygen and regulate temperature.

Preservation

Preservation techniques are needed to prevent fish spoilage and lengthen shelf life. They are designed to inhibit the activity of spoilage bacteria and the metabolic changes that result in the loss of fish quality. Spoilage bacteria are the specific bacteria that produce the unpleasant odours and flavours associated with spoiled fish. Fish normally host many bacteria that are not spoilage bacteria, and most of the bacteria present on spoiled fish played no role in the spoilage. To flourish, bacteria need the right temperature, sufficient water and oxygen, and surroundings that are not too acidic. Preservation techniques work by interrupting one or more of these needs. Preservation techniques can be classified as follows.

Control of Temperature

If the temperature is decreased, the metabolic activity in the fish from microbial or autolytic processes can be reduced or stopped. This is achieved by refrigeration where the temperature is dropped to about 0 °C, or freezing where the temperature is dropped below -18 °C. On fishing vessels, the fish are refrigerated mechanically by circulating cold air or by packing the fish in boxes with ice. Forage fish, which are often caught in large numbers, are usually chilled with refrigerated or chilled seawater. Once chilled or frozen, the fish need further cooling to maintain the low temperature. There are key issues with fish cold store design and management, such as how large and energy efficient they are, and the way they are insulated and palletized.

Ice preserves fish and extends shelf life by lowering the temperature

An effective method of preserving the freshness of fish is to chill with ice by distributing ice uniformly around the fish. It is a safe cooling method that keeps the fish moist and in an easily stored form suitable for transport. It has become widely used since the development of mechanical refrigeration, which makes ice easy and cheap to produce. Ice is produced in various shapes; crushed ice and ice flakes, plates, tubes and blocks are commonly used to cool fish. Particularly effective is slurry ice, made from micro crystals of ice formed and suspended within a solution of water and a freezing point depressant, such as common salt.

A more recent development is pumpable ice technology. Pumpable ice flows like water, and because it is homogeneous, it cools fish faster than fresh water solid ice methods and eliminates freeze burns. It complies with HACCP and ISO food safety and public health standards, and uses less energy than conventional fresh water solid ice technologies.

Ice manufactured in this ice house is delivered down the Archimedes screw into the ice hold on the boat, Pittenweem

Control of Water Activity

The water activity, a_w, in a fish is defined as the ratio of the water vapour pressure in the flesh of the fish to the vapour pressure of pure water at the same temperature and pressure. It ranges between 0 and 1, and is a parameter that measures how available the water is in the flesh of the fish. Available water is necessary for the microbial and enzymatic reactions involved in spoilage. There are a number of techniques that have been or are used to tie up the available water or remove it by reducing the a_w. Traditionally, techniques such as drying, salting and smoking have been used, and have been used for thousands of years. These techniques can be very simple, for example, by using solar drying. In more recent times, freeze-drying, water binding humectants, and fully automated equipment with temperature and humidity control have been added. Often a combination of these techniques is used.

Women drying fish, 1971

Dry fish market at Mohanganj

Drying stockfish in Iceland

Fish barn with fish drying in the sun – Van Gogh 1882.

Physical Control of Microbial Loads

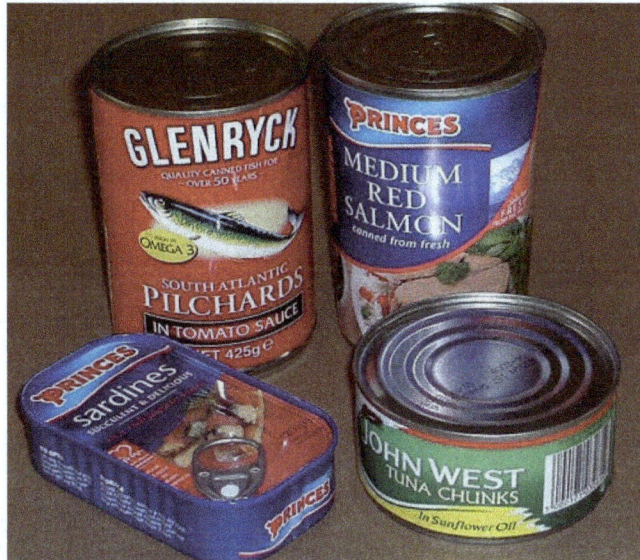

Microbial loads can be physically controlled by canning and then sterilizing with heat

Heat or ionizing irradiation can be used to kill the bacteria that cause decomposition. Heat is applied by cooking, blanching or microwave heating in a manner that pasteurizes or sterilizes fish products. Cooking or pasteurizing does not completely inactivate microorganisms and may need to be followed with refrigeration to preserve fish products and increase their shelf life. Sterilised products are stable at ambient temperatures up to 40 °C, but to ensure they remain sterilized they need packaging in metal cans or retortable pouches before the heat treatment.

Chemical Control of Microbial Loads

Microbial growth and proliferation can be inhibited by a technique called biopreservation. Biopreservation is achieved by adding antimicrobials or by increasing the acidity of the fish muscle. Most bacteria stop multiplying when the pH is less than 4.5. Acidity is increased by fermentation, marination or by directly adding acids (acetic, citric, lactic) to fish products. Lactic acid bacteria produce the antimicrobial nisin which further enhances preservation. Other preservatives include nitrites, sulphites, sorbates, benzoates and essential oils.

Control of The Oxygen Reduction Potential

Spoilage bacteria and lipid oxidation usually need oxygen, so reducing the oxygen around fish can increase shelf life. This is done by controlling or modifying the atmosphere around the fish, or by vacuum packaging. Controlled or modified atmospheres have specific combinations of oxygen, carbon dioxide and nitrogen, and the method is often combined with refrigeration for more effective fish preservation.

Combined Techniques

Two or more of these techniques are often combined. This can improve preservation and reduce unwanted side effects such as the denaturation of nutrients by severe heat treatments. Common combinations are salting/drying, salting/marinating, salting/smoking, drying/smoking, pasteur-ization/refrigeration and controlled atmosphere/refrigeration. Other process combinations are currently being developed along the multiple hurdle theory.

Automated Processes

"The search for higher productivity and the increase of labor cost has driven the development of computer vision technology, electronic scales and automatic skinning and filleting machines."

Fish feed production in Norway

Waste Management

Non edible fish scrap processing, 1884

Waste produced during fish processing operations can be solid or liquid.

- Solid wastes: include skin, viscera, fish heads and carcasses (fish bones). Solid waste can be recycled in fish meal plants or it can be treated as municipal waste.

- Liquid wastes: include bloodwater and brine from drained storage tanks, and water dis-charges from washing and cleaning. This waste may need holding temporarily, and should be disposed of without damage to the environment. How liquid waste should be disposed from fish processing operations depends on the content levels in the waste of solid and organic matter, as well as nitrogen and phosphorus content, and oil and grease content. It also depends on an assessment of parameters such acidity levels, temperature, odour, and biochemical oxygen demand and chemical oxygen demand. The magnitude of waste man-agement issues depends on how much waste volume there is, the nature of the pollutants it carries, the rate at which it is discharged and the capacity of the receiving environment to assimilate the pollutants. Many countries dispose of such liquid wastes through their municipal sewage systems or directly into a waterway. The receiving waterbody should be able to degrade the organic and inorganic waste components in a way that does not damage the aquatic ecosystem.

Treatments can be primary and secondary.

- Primary treatments: use physical methods such as flotation, screening, and sedimentation to remove oil and grease and other suspended solids.

- Secondary treatments: use biological and physicochemical means. Biological treatments use microorganisms to metabolise the organic polluting matter into energy and biomass. "These microorganisms can be aerobic or anaerobic. The most used aerobic processes are activated sludge system, aerated lagoons, trickling filters or bacterial beds and the rotating biological contractors. In anaerobic processes, the anaerobic microorganisms digest the organic matter in tanks to produce gases (mainly methane and CO_2) and biomass. Anaerobic digesters are sometimes heated, using part of the methane produced, to maintain a temperature of 30 to 35°C. In the physicochemical treatments, also called coagulation-flocculation, a chemical substance is added to the effluent to reduce the surface charges responsible for particle repulsions in a colloidal suspension, thus reducing the forces that keep its particles apart. This reduction in charge causes flocculation (agglomeration) and particles of larger sizes are settled and clarified effluent is obtained. The sludge produced by primary and secondary treatments is further processed in digesting tanks through anaerobic processes or sprayed over land as a fertilizer. In the latter case, care must be exercised to ensure that the sludge is freed of its pathogens."

Transport

Fish is transported widely in ships, and by land and air, and much fish is traded internationally. It is traded live, fresh, frozen, cured and canned. Live, fresh and frozen fish need special care.

- Live fish: When live fish are transported they need oxygen, and the carbon dioxide and ammonia that result from respiration must not be allowed to build up. Most fish transported live are placed in water supersaturated with oxygen (though catfish can breathe air directly through their gills and body skin, and the climbing perch has special air-breathing organs). The fish are often "conditioned" (starved) before they are transported to reduce their metabolism and increase packing density, and the water can be cooled to further reduce metabolism. Live crustaceans can be packed in wet sawdust to keep the air humid.

- By air: Over five percent of the global fish production is transported by air. Air transport needs special care in preparation and handling and careful scheduling. Airline transport hubs often require cargo transfers under their own tight schedules. This can influence when the product is delivered, and consequently the condition it is in when it is delivered. The air shipment of leaking seafood packages causes corrosion damage to aircraft, and each year, in the US, requires millions of dollars to repair the damage. Most airlines prefer fish that is packed in dry ice or gel, and not packed in ice.

- By land or sea: "The most challenging aspect of fish transportation by sea or by road is the maintenance of the cold chain, for fresh, chilled and frozen products and the optimisation of the packing and stowage density. Maintaining the cold chain requires the use of insulated containers or transport vehicles and adequate quantities of coolants or mechanical refrigeration. Continuous temperature monitors are used to provide evidence that the cold chain has not been broken during transportation. Excellent development in food packaging and handling allow rapid and efficient loading, transport and unloading of fish and fishery products by road or by sea. Also, transport of fish by sea allows for the use of special containers that carry fish under vacuum, modified or controlled atmosphere, combined with refrigeration."

Quality and Safety

The International Organization for Standardisation, ISO, is the worldwide federation of national standards bodies. ISO defines *quality* as "the totality of features and characteristics of a product or service that bear on its ability to satisfy stated or implied needs."(ISO 8402). The quality of fish and fish products depends on safe and hygienic practices. Outbreaks of fish-borne illnesses are reduced if appropriate practices are followed when handling, manufacturing, refrigerating and transporting fish and fish products. Ensuring standards of quality and safety are high also minimizes the post-harvest losses."

"The fishing industry must ensure that their fish handling, processing and transportation facilities meet requisite standards. Adequate training of both industry and control authority staff must be provided by support institutions, and channels for feedback from consumers established. Ensuring high standards for quality and safety is good economics, minimizing losses that result from spoilage, damage to trade and from illness among consumers."

Fish processing highly involves very strict controls and measurements in order to ensure that all processing stages have been carried out hygienically. Thus, all fish processing companies are highly recommended to join a certain type of food safety system. One of the certifications that are commonly known is the Hazard Analysis Critical Control Points (HACCP).

Hazard Analysis and Critical Control Points

HACCP is a system which identifies hazards and implements measures for their control. It was first developed in 1960 by NASA to ensure food safety for the manned space program. The main objectives of NASA were to prevent food safety problems and control food borne diseases. HACCP has been widely used by food industry since the late 1970 and now it is internationally recognized as the best system for ensuring food safety.

"The Hazard Analysis and Critical Control Points (HACCP) system of assuring food safety and quality has now gained worldwide recognition as the most cost-effective and reliable system available. It is based on the identification of risks, minimizing those risks through the design and layout of the physical environment in which high standards of hygiene can be assured, sets measurable standards and establishes monitoring systems. HACCP also establishes procedures for verifying that the system is working effectively. HACCP is a sufficiently flexible system to be successfully applied at all critical stages -- from harvesting of fish to reaching the consumer. For such a system to work successfully, all stakeholders must cooperate which entails increasing the national capacity for introducing and maintaining HACCP measures. The system's control authority needs to design and implement the system, ensuring that monitoring and corrective measures are put in place."

HACCP is endorsed by the:

- FAO (Food and Agriculture Organization)

- Codex Alimentarius (a commission of the United Nations)

- FDA (US Food and Drug Administration)

- European Union

- WHO (World Health Organization)

There are seven basic principles:

- Principle 1: Conduct a hazard analysis.

- Principle 2: After assessing all the processing steps, the Critical control point (CCP) is controlled. CCP are points which determine and control significant hazards in a food manufacturing process.

- Principle 3: Set up critical limits in order to ensure that the hazard identified is being controlled effectively.

- Principle 4: Establish a system so as to monitor the CCP.

- Principle 5: Establish corrective actions where the critical limit has not been met. Appropriate actions need to be taken which can be on a short or long-term basis. All records must be sustained accurately.

- Principle 6: Establish authentication procedures so as to confirm if the principles imposed by HACCP documents are being respected effectively and all records are being taken.

- Principle 7: Analyze if the HACCP plan are working effectively.

Final Products

Finfish, or parts of finfish, are typically presented physically for marketing in one of the following forms

- whole fish: the fish as it originally came from the water, with no physical processing

- drawn fish: a whole fish which has been eviscerated, that is, had its internal organs removed

- dressed fish: fish that has been scaled and eviscerated, and is ready to cook.

- pan dressed fish: a dressed fish which has had its head, tail, and fins removed, so it will fit in a pan.

- filleted fish: the "fleshy sides of the fish, cut lengthwise from the fish along the backbone. They are usually boneless, although in some fish small bones called "pins" may be present; skin may be present on one side, too. Butterfly fillets may be available. This refers to two fillets held together by the uncut flesh and skin of the belly"

- fish steaks: large dressed fish can be cut into cross section slices, usually half to one inch thick, and usually with a cross section of the backbone

- fish sticks: "are pieces of fish cut from blocks of frozen fillets into portions at least 3/8-inch thick. Sticks are available in fried form ready to heat or frozen raw, coated with batter and breaded, ready to be cooked"

- fish cakes: are "prepared from flaked fish, potatoes, and seasonings, and shaped into cakes, coated with batter, breaded, and then packaged and frozen, ready-to-be-cooked"

- fish fingers

Value Addition

Imitation crab and imitation shrimp made from surimi

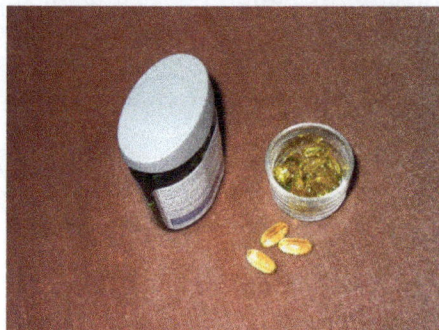

Fish oil capsules

In general value addition means "any additional activity that in one way or the other change the nature of a product thus adding to its value at the time of sale." Value addition is an expanding sector in the food processing industry, especially in export markets. Value is added to fish and fishery products depending on the requirement of different markets. Globally a transition period is taking place where cooked products are replacing traditional raw products in consumer preference.

"In addition to preservation, fish can be industrially processed into a wide array of products to increase their economic value and allow the fishing industry and exporting countries to reap the full benefits of their aquatic resources. In addition, value processes generate further employment and hard currency earnings. This is more important nowadays because of societal changes that have led to the development of outdoor catering, convenience products and food services requiring fish products ready to eat or requiring little preparation before serving."

"However, despite the availability of technology, careful consideration should be given to the economic feasibility aspects, including distribution, marketing, quality assurance and trade barriers, before embarking on a value addition fish process."

- Surimi: Surimi and surimi-based products are an example of value added products. Surimi is prepared from the mechanically deboned, washed (bleached) and stabilised flesh of fish. "It is an intermediate product used in the preparation of a variety of ready to eat seafood such as kamaboko, fish sausage, crab legs and imitation shrimp products. Surimi-based products are gaining more prominence worldwide, because of the emergence of Japanese restaurants and culinary traditions in North America, Europe and elsewhere. Ideally, surimi should be made from low-value, white fish with excellent gelling ability and which are abundant and available year-round. At present, Alaskan pollack accounts for a large proportion of the surimi supply. Other species, such as sardine, mackerel, barracuda, striped mullet have been successfully used for surimi production."

- Fishmeal and fish oil: "A significant proportion of the world catch (20 percent) is processed into fishmeal and fish oil. Fishmeal is a ground solid product that is obtained by removing most of the water and some or all of the oil from fish or fish waste. This industry was launched in the 19th century, based mainly on surplus catches of herring from seasonal coastal fisheries to produce oil for industrial uses in leather tanning and in the production of soap, glycerol and other non-food products. Presently, it uses small oily fish to produce fishmeal and oil. It is worthy to mention that, only where it is uneconomic or impracticable for human consumption, should the catch be reduced to fishmeal and oil. Indeed, cycling fish through poultry or pigs is a loss because there is a need for 3 kg of edible fish to produce approximately 1 kg of edible chicken or pork."

History

Egyptians bringing in fish and splitting them for salting

A medieval view of fish processing, by Peter Brueghel the Elder (1556).

There is evidence humans have been processing fish since the early Holocene. For example, fish-bones (c. 8140–7550 BP, uncalibrated) at Atlit-Yam, a submerged Neolithic site off Israel, have been analysed. What emerged was a picture of "a pile of fish gutted and processed in a size-dependent manner, and then stored for future consumption or trade. This scenario suggests that technology for fish storage was already available, and that the Atlit-Yam inhabitants could enjoy the economic stability resulting from food storage and trade with mainland sites."

Medieval smokehouse at Walraversijde, ca. 1465

Ice house used to preserve fish at Findhorn

References

- Béné, C; Macfadyen, G; Allison, E H (2007) Increasing the contribution of small-scale fisheries to poverty alleviation and food security FAO Fisheries Technical Paper T481. ISBN 978-92-5-105664-6

- Bekker-Nielsen T (2005) Ancient fishing and fish processing in the Black Sea region Volume 2 of Black Sea studies, Aarhus University Press, ISBN 978-87-7934-096-1.

- Brewer DJ and Friedman RF (1989) Fish and Fishing in Ancient Egypt Cairo press: The American University in Cairo. ISBN 978-977-424-224-3

- Pearson AM and Dutson TR (1999) HACCP in Meat, Poultry and Fish Processing, Volume 10 of Advances in meat research, Springer. ISBN 978-0-8342-1327-2.

- Stellman JM (ed.) (1998) Chemical, industries and occupations Volume 3 of Encyclopaedia of Occupational Health and Safety, International Labour Organization. ISBN 978-92-2-109816-4.

- Stewart H (1982) Indian Fishing: Early Methods on the Northwest Coast University of Washington Press. ISBN 978-0-88894-332-3.

- United Nations Development Fund for Women (1993) Fish processing Food Technology Source Book Series (UNIFEM) Series, ISBN 978-1-85339-137-8.

- "In Mackerel's Plunder, Hints of Epic Fish Collapse". International Herald Tribune. 25 January 2012. Retrieved 30 January 2016 – via The New York Times.

- Krah, Jaclyn; Unger, Richard L. (7 August 2013). "The Importance of Occupational Safety and Health: Making for a "Super" Workplace". National Institute for Occupational Safety and Health. Retrieved 16 January 2015.

- "Commercial Fishing Safety". NIOSH Workplace Safety and Health Topics. National Institute of Occupational Safety and Health. Retrieved 11 July 2012.

- Welch, Laine (January 30, 2010). "Device makes fishing deck winch less lethal". Anchorage Daily News. Retrieved 11 July 2012.

- Davis, Mary E. "Occupational Safety and Regulatory Compliance in US Commercial Fishing". Archives of Environmental & Occupational Health. 66 (4): 209–216. doi:10.1080/19338244.2011.564237.

- A Mood and P Brooke (July 2010). Estimating the Number of Fish Caught in Global Fishing Each Year. Fish-Count.org.uk.

Fishing Techniques and Methods

This chapter studies the array of techniques that are employed for fishing globally and some of these include longline fishing, castnet, jug fishing, surf fishing and trawling. The topics listed combine simple and basic technologies with modern methods. This chapter discusses the methods of fishing in a critical manner providing key analysis to the subject matter.

Longline Fishing

Longlining for mackerel

Longline fishing is a commercial fishing technique. It uses a long line, called the main line, with baited hooks attached at intervals by means of branch lines called *snoods* (or *gangions*). A snood is a short length of line, attached to the main line using a clip or swivel, with the hook at the other end. Longlines are classified mainly by where they are placed in the water column. This can be at the surface or at the bottom. Lines can also be set by means of an anchor, or left to drift. Hundreds or even thousands of baited hooks can hang from a single line. Longliners commonly target swordfish, tuna, halibut, sablefish and many other species.

In some unstable fisheries, such as the Patagonian toothfish, fishermen may be limited to as few as 25 hooks per line. In contrast, commercial longliners in certain robust fisheries of the Bering Sea and North Pacific generally run over 2,500 hand-baited hooks on a single series of connected lines many miles in length.

Longline radiobuoys

Longlines can be set to hang near the surface (pelagic longline) to catch fish such as tuna and swordfish or along the sea floor (demersal longline) for groundfish such as halibut or cod. Longliners fishing for sablefish, also referred to as black cod, occasionally set gear on the sea floor at depths exceeding 1,100 metres (3,600 ft) using relatively simple equipment. Longlines with traps attached rather than hooks can be used for crab fishing in deep waters.

Longline fishing is prone to the incidental catching and killing of seabirds, sea turtles, and sharks, but can be considerably more ecologically sustainable than some other commercially significant harvesting methods.

Incidental Catch

Black-browed albatross hooked on a long-line

Longline fishing is controversial in some areas because of bycatch, fish caught while seeking another species or immature juveniles of the target species. This can cause many issues, such as the killing of many other marine animals while seeking certain commercial fish. Seabirds can be particularly vulnerable during the setting of the line.

Methods to mitigate incidental mortality have succeeded in some fisheries. Mitigation techniques include the use of weights to ensure the lines sink quickly, the deployment of streamer lines to scare away birds, setting lines only at night in low light (to avoid attracting birds), limiting fishing seasons to the southern winter (when most seabirds are not feeding young), and not discharging offal while setting lines.

The Hawaii-based longline fishery for swordfish was closed in 2000 over concerns of excessive sea turtle by-catch, particularly loggerhead sea turtles and leatherback turtles. Changes to the management rules allowed the fishery to reopen in 2004. Gear modification, particularly a change to large circle-hooks and mackerel-type baits, eliminated much of the sea turtle by-catch associated with the fishing technique. It has been claimed that one consequence of the closure was that 70 Hawaii-based vessels were replaced by 1,500-1,700 longline vessels from various Asian nations, but this is not based on any reliable data. Due to poor and often non-existent catch documentation by these vessels, the number of sea turtles and albatross caught by these vessels between 2000 and 2004 will never be known. Hawaii longline fishing for swordfish closed again on 17 March 2006, when the by-catch limit of 17 loggerhead turtles was reached. In 2010 the by-catch limit for loggerhead turtles was raised, but was restored to the former limit as a result of litigation. The Hawaii-based longline fisheries for tuna and swordfish are managed under sets of slightly different rules. The tuna fishery is one of the best managed fisheries in the world, according to the UN Code of Responsible Fishing, but has been criticized by others as being responsible for continuing by-catch of false killer whales, seabirds, and other nontargeted wildlife, as well as placing pressure on depleted bigeye tuna stocks.

Commercial longline fishing is also one of the main threats to albatrosses. Of the 22 albatross species recognized in the IUCN Red List, six are threatened, and nine are vulnerable. The IUCN lists three species as critically endangered: the Amsterdam albatross, the Tristan albatross and the waved albatross. The remaining four are near threatened. Albatrosses and other seabirds which readily feed on offal are attracted to the set bait, become hooked on the lines and drown. An estimated 100,000 albatross per year are killed in this way.

Safety

Researchers found that the annual number of injuries in the freezer longline fleet was 25 per year during 2001-2012 for a total of 303 work-related injuries (9 fatal, 294 non-fatal). The risk for non-fatal injuries was 35 per 1,000 full-time equivalent employees, about three times higher than average U.S. worker.

Historic Images

Commercial Longlining

Preparing lines for longlining

Snoods (gangions) used in the longlining

Cast Net

A fisherman casting a net in Kerala, India

A cast net, also called a throw net, is a net used for fishing. It is a circular net with small weights distributed around its edge.

The net is cast or thrown by hand in such a manner that it spreads out on the water and sinks. This technique is called net casting or net throwing. Fish are caught as the net is hauled back in. This simple device is particularly effective for catching small bait or forage fish, and has been in use, with various modifications, for thousands of years. On the US Gulf Coast, it is used especially to catch mullet, which will not bite a baited hook.

Construction and Technique

Contemporary cast nets have a radius which ranges from 4 to 12 feet (1.2 to 3.6 metres). Only strong people can lift the larger nets once they are filled with fish. Standard nets for recreational fishing have a four-foot hoop. Weights are usually distributed around the edge at about one pound per foot (1.5 kilograms per metre). Attached to the net is a handline, one end of which is held in the hand as the net is thrown. When the net is full, a retrieval clamp, which works like a wringer on a mop, closes the net around the fish. The net is then retrieved by pulling on this handline. The net is lifted into a bucket and the clamp is released, dumping the caught fish into the bucket.

Cast nets work best in water no deeper than their radius. Casting is best done in waters free of obstructions. Reeds cause tangles and branches can rip nets. The net caster may choose to stand with one hand holding the handline, and with the net draped over the other arm so that the weights dangle, or, with most of the net being held in one hand and only a part of the lead line held in the other hand so the weights dangle in a staggered fashion (approximately half of the weights in the throwing hand being held higher than the rest of the weights). The line is then thrown out to the water, using both hands, in a circular motion rather as in hammer throwing. The net can be cast from a boat, or from the shore, or by wading.

There are also optional net throwers that can make casting easier. These look like a lid from a trash can, including the handle on top. The outside circumference has a deep gutter. The net is loaded along the gutter and the weights are placed inside the gutter. The net is then tossed into the water using the thrower.

Mosaic, 4th century BC, showing a retiarius or "net fighter", with a trident and cast net, fighting a secutor.

In Norse mythology the sea giantess Rán cast a fishing net to trap lost sailors. In Ancient Rome, in a parody of fishing, a type of gladiator called a retiarius or "net fighter" was armed with a trident and a cast net. The retiarius was traditionally pitted against a secutor.

Between 177 and 180 the Greek author Oppian wrote the *Halieutica*, a didactic poem about fishing. He described various means of fishing including the use of nets cast from boats. References to cast nets can also be found in the New Testament.

Jug Fishing

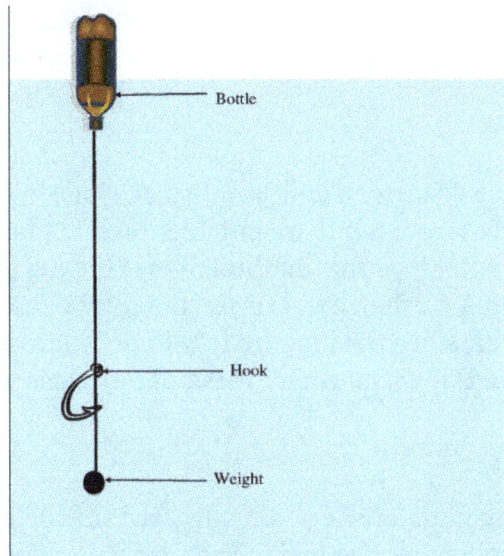

Jug Fishing Image

Jug fishing is an unlimited class tackle method of fishing that uses lines suspended from floating jugs to catch fish in lakes or rivers. Often, a large number of jugs are used when jug fishing. In many states a fisherman could use up to twenty, and jug sets of around twenty are common in practice. Jug fishing is most common in southern states where many different kinds of people jug fish. Jugs are often put out at sunset and picked up at sunrise by the whole family. Jug fishing consists of a simple setup where lines are tied onto jugs and weights can be added to the line to keep the jug's location fixed. Jug fishing is also subjected to numerous Department of Natural Resources and local water regulations that could include: the number of jugs, dates and times when jug fishing is allowed, and if jug fishing is even allowed. Many fish can be caught on jugs, but the main target of jug fishing is often catfish.

Regulations

Jug fishing is not permitted on certain waters throughout the United States. Before jug fishing, a fisherman should check with each water's regulations to see if jug fishing is allowed. Also, each water might have different regulations regarding jug fishing when compared to the regulations of the Department of Natural Resources.

DNR Regulations

According to DNR regulations, each jug must be free-floating. The jug's size must be at least one pint but no more than one gallon. Only one line may be attached to each jug. In order to fish with jugs, a permit is also required in some areas. The maximum number of jugs that is allowed to be used by one individual is 50. All jugs must also be marked with the individual's name and address.

Jug fishing is also limited to certain hours during the day. Jug may only be used up to one hour after official sunrise and can be placed back on the water no earlier than one hour before official sunset.

Techniques

On average, a fisherman will use twenty or more jugs when fishing. One common set up for these jugs is to use a two-liter bottle that has lines, hooks, weights, and bait attached to it. Two main strategies are usually employed when jug fishing, free floating jugs and fixed jugs.

Free Floating Jugs

When fishing free floating jugs, a fisherman will simply place jugs in the water that have hooks and lines attached to them, they have weights but are not anchored in place. The jug will then be free to move about in the water's current. Since this method allows the jug to drift through a large area of water, this method is useful when a fisherman is uncertain of the location of the fish. This method is very simple to set up. However, since the jugs are free to drift across the lake, the jugs are easy to lose (especially if a fish is hooked) if a fisherman does not pay close attention to the jugs.

Fixed Floating Jugs

When fishing with fixed floating jugs, a fisherman will place the jug in one location and fix the jug to that location by one of the following two ways. The first is by tying the jug to a branch, stump, or another fixed object on the water. The second way is by attaching large weights (approximately one to three pounds depending on the current in the water) to the bottom of the fishing line below the hook to keep the jug from moving. This method is particularly effective for catching large fish. Also, the jug will not become lost since it is fixed. However, since the jug is fixed, this now requires the fish to find the jug which is always an uncertainty.

Guidelines

After the jugs have been placed in the water, a person will usually maneuver about the water in a boat to keep sight of the jugs. A jug will usually bob and weave in the water when a fish has been hooked. Jugs are usually baited with but not limited to shad, small fish, and nightcrawlers.

Controversy

Jug fishing is a source of controversy. Many people think that people who jug fish are over-harvesting the fish and not promoting a successful fishing environment for the future. However, each water has its own regulations concerning how many fish can be caught and how the fish can be caught. These regulations are designed to protect fish and provide a sustainable environment for the fish.

Surf Fishing

Surf fishing is the sport of catching fish standing on the shoreline or wading in the surf. A general term, surf fishing may or may not include casting a lure or bait, and refers to all types of shore fishing - from sandy and rocky beaches, rock jetties, or even fishing piers. The terms surfcasting or beachcasting refer more specifically to surf fishing from the beach by casting into the surf at or near the shoreline. With few exceptions, surf fishing is done in saltwater.

Surf Fisherman

Equipment

Surf fishing Southampton

The basic idea of most surfcasting is to cast a bait or lure as far out into the water as is necessary to reach the target fish from the shore. This may or may not require long casting distances. Basic surf fishing can be done with a fishing rod between 7 feet and 18 feet long, with an extended butt section, equipped with a suitably-sized spinning or conventional (revolving spool) casting reel. In addition to rod and reel, the surf fisherman needs terminal tackle and bait or lure. Terminal tackle is the equipment at the far end of the line: hooks, swivels, lines and leaders.

Dedicated surfcasters usually possess an array of terminal and other tackle, with fishing rods and reels of different lengths and actions, and lures and baits of different weights and capabilities. Depending on fishing conditions and the type of fish they are trying to catch, such surfcasters tailor bait and terminal tackle to rod and reel and the size and species of fish targeted. Reels and other equipment need to be constructed so they resist the corrosive and abrasive effects of salt and sand.

Surf fishermen who use artificial lures, cast and retrieve them to entice a bite from the target species. There are hundreds of different lures effective for surf fishing, such as spoons, plugs, soft plastics and jigs. Most can be purchased from local bait and tackle shops, online tackle retailers, at fishing tackle expositions or specialized surf fishing catalogs. Most surfcasters carry with them a "surf bag" which holds a selection of lures to facilitate fast changes of lures appropriate to current fishing conditions, saving trips back to the beach or vehicle to change equipment.

Several other items of equipment are commonly used by surf fisherman and surfcasters to improve comfort, convenience and effectiveness. Among these are waders, used to wade out into the surf to gain distance from shore when casting the bait. Full length, chest-high waders are most popular, in order to provide a measure of protection against a pounding surf that could easily swamp hip-length wading boots. In addition to the extra reach provided by wading out from shore, waders provide improved footing, protection for feet and legs from sharp bottom objects and stinging/biting fish and crustaceans, and protection from cold water temperatures. Most surf fishermen prefer integrated booted waders to stocking-foot models, which eliminates the tendency of sand and rock to find their way in between boot and wader. In areas where the bottom consists of slippery rocks or when fishing from mossy and slimy rock jetties, cleated boots or attachments are utilized to improve footing and enhance safety.

Surf fishing is done often at night to follow the nocturnal feeding habits of many target species. Many surf fishermen add items such as flashlights, headlamps, light sticks and other gear to facilitate night fishing.

Surfcasting

Surfcasting is a casting technique which separates the surfcaster from the ordinary shore, pier, or boat fisherman. Extremely long rods are frequently employed to extend the length of the cast, while specialized, two-handed casting techniques are used to cast the lure or bait the added distances required in many cases to reach feeding inshore fish. In these casts the entire body, rather than just the arms, are utilized to deliver the cast. In addition to standard two-handed casts, veteran surfcasters may also resort to the *pendulum cast* (derived from tournament casting contests) to achieve added distance - in some cases exceeding 700 feet.

Tournament casting is a sport in its own right, with the world record held by Danny Moeskops casting a distance of 286.63 meters (313.46 yards).

Extreme Surfcasting also known as Skishing is a variant of surf fishing that involves donning a wetsuit and flippers and swimming into the ocean with surfcasting pole to catch fish. Skishing was invented in Montauk, NY by local resident, Paul Melnyk, in 1995. He coined the term which is a cross between "skiing "and "fishing" to describe what happens if the angler hooks a large enough fish and it tows him through the ocean like a water-skier. Without the benefit of boat or land, the fight is considered to be, by the fisherman anyway, on more equal terms. Skishers swim sometimes hundreds of yards from shore to water well over their heads, with their flippers and buoyancy of their wetsuits keeping them afloat.

Beachcasting

In Britain, surfcasting is often called beachcasting. It is a popular form of fishing which is carried out all around the coast of the British Isles. Beachcasters use very long fishing rods, usually between 12 and 16 feet. The beachcaster will stand on a beach or shoreline and cast out to sea with either a water filled float, or a lead weight weighing between 120g and 200g. Bait used in this form of fishing might include limpets, mussels, lugworm, ragworm, sandeel, mackerel strip, squid, peeler crab or razor fish. Additionally, artificial flies or even spinners may be used for species such as mackerel or bass. It is a common pastime in coastal areas of Scotland, England, Wales and Ireland and often results in the capture of large specimens of many species of fish, including: flatfish, bass, cod, whiting, pollack, black bream, dogfish, smooth hound, bull huss, rays and tope.

Common beachcasting techniques used in Britain include the common overhead cast, the off-the-ground Cast and the Pendulum cast. There are other techniques used in Britain, but these three techniques remain the more popular ones with the pendulum cast perhaps the most difficult to master and also the one that generally produces more yards to the cast with 250-300 yds becoming quite possible. On top of technique and equipment, streamlines rigs that can take casting strains are also used such as the clipped pennel pulley, clipped down rigs, long n low etc. The rigs are also made with strong 60 lb plus lines to take the strain of the cast. Streamlining of bait too is important. A shockleader is a stronger line attached to a lighter mainline to absorb the shock of a powerful cast. The suggested formula for shockleader selection is as calculated as follows 1 oz weight +10lb shockleader + 10lb, e.g.

Besides the equipment and the cast techniques, the other component that makes a beach caster successful is what can be called sea interpretation. There are zones, and within them certain points, where to cast the baits is more profitable than others. It depends on the conformation of the sea bed, and therefore, in order to make fishing more effective and enjoyable, we should spend a while on understanding these differences and only then start fishing. How? First of all, by observing the waves motion, in particular where they break out, because it is there where the bottom height is lower and the sea current can stir it most. This action uncovers from the sandy bed small marine beings and carry them away across channels to deeper points, where they settle in for the joy of fish. The latter, that tend to swim countercurrent, move searching for food along these channels - that can be compared in some way to our roads in a map - and approach the shore as long as the conditions allow the marine food chain to last. Although it may seem an oversimplification of the reality, it makes clear that only when a beach caster is able to master the "sense of sea" by spotting the difference among fishing zones, then will he be more likely to make memorable captures. Similarly to any other activity, it's a matter of experience, which can be acquire by getting through a several number of failures.

Dangers

As with any water sport care must be taken to participate safely in this sport. Much surf fishing is done in rough whitewater surf conditions. Powerful waves and strong undertows can cause serious injury or death if proper attention is not paid to safety.

Areas fished should be scouted in low tide conditions to note sudden drop offs or dangerous conditions hidden at high tides. Any fisherman with waders should wear a wader belt to keep waders from filling with water in the event one falls in the surf. The aforementioned cleats should be worn anywhere there are slippery rocks or shells underfoot. PFD's (personal flotation devices) should be considered especially when fishing alone in big surf or on jetties.

Since lures and hooks feature razor sharp points, care must be taken not to hook oneself or others when casting, especially when performing two-handed full power casts that require a substantial safety zone behind the surfcaster. In the event that one accidentally hooks oneself or someone else, it is a good idea to carry a quality cutting pliers capable of cutting the hooks you are fishing with.

Species

A wide array of species can be targeted from surf and shore limited only the species of fish available in the region. In the eastern coast of the United States, the striped bass is highly valued. This species can be fished from shore and ranges in weight from a few pounds to the world record 78.5 lb (35.6 kg). Fish in the 30 to 40 lb (15 kg) range are common on the mid-Atlantic US coast from New York to the Carolinas. Some other available species are bluefish, redfish (red drum), black drum, tautog (blackfish), flounder (fluke), weakfish (sea trout), bonito and albacore tuna, pompano, Spanish mackerel, snook and tarpon. Even sharks can be targeted by surf fishermen.

From North Carolina south the redfish (red drum) is one of the most targeted fish by surf anglers. Fishermen use rods 10 to 13 feet to cast to drum using baits like cut mullet, bunker (cut menhaden, or pogies as Carolina anglers call them), and cut bait from spot, croaker, or bluefish. Red drum hit a bait aggressively and are a much beloved surf fishing species. The overfished status of this fish for many years due to the blackened redfish craze of the 1980s led to strict recreational size and creel limits, however, so surf anglers must learn the local rules.

Beach buggies

Many areas allow four-wheel-drive (4WD) vehicles on to the beach. This allows the surf fisherman to scout and fish large stretches of shoreline. Although the term "beach buggy" may be applied to special vehicles for transportation on sand, 4x4 trucks and SUV's with deflated tires are more often used in surf fishing. Permits are usually required and need to be obtained from the appropriate state or local authorities. Most require an additional list of safety and other equipment, sometimes called Beach Permit Kits to ensure that the vehicle and its inhabitants can safely navigate the soft sand and are prepared in the event the vehicle gets stuck. Beach Buggy access can be hindered at times by beach closures, due to endangered bird species nesting. Beach buggy access is often hotly contested between environmental groups, and beach access enthusiasts. Therefore it is a good idea to check on local regulations before you attempt to drive your vehicle on the beach. Driving in restricted areas can result in serious penalties.

Trawling

Trawling is a method of fishing that involves pulling a fishing net through the water behind one or more boats. The net that is used for trawling is called a trawl.

The boats that are used for trawling are called trawlers or draggers. Trawlers vary in size from small open boats with as little as 30 hp (22 kW) engines to large factory trawlers with over 10,000 hp (7.5 MW). Trawling can be carried out by one trawler or by two trawlers fishing cooperatively (pair trawling).

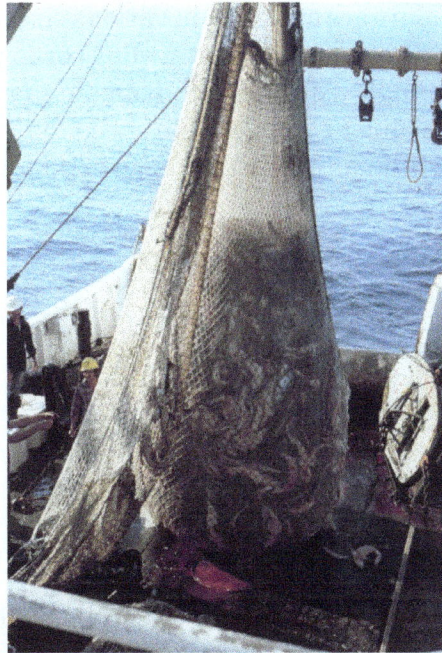

Trawl net with fish

Trawling can be contrasted with trolling, where baited fishing lines instead of trawls are drawn through the water. Trolling is used both for recreational and commercial fishing whereas trawling is used mainly for commercial fishing. Trawling is also commonly used as a scientific sampling, or survey, method.

Bottom Versus Midwater Trawling

Trawl catch of myctophids and glass shrimp from the bottom at greater than 200 meters depth

Trawling can be divided into bottom trawling and midwater trawling, depending on how high the trawl (net) is in the water column. Bottom trawling is towing the trawl along (benthic trawling) or close to (demersal trawling) the sea floor. Midwater trawling is towing the trawl through free water above the bottom of the ocean or benthic zone.

Midwater trawling is also known as pelagic trawling. Midwater trawling catches pelagic fish such as anchovies, shrimp, tuna and mackerel, whereas bottom trawling targets both bottom-living fish (groundfish) and semi-pelagic fish such as cod, squid, halibut and rockfish.

The gear itself can vary a great deal. Pelagic trawls are typically much larger than bottom trawls, with very large mesh openings in the net, little or no ground gear, and little or no chaffing gear. Additionally, pelagic trawl doors have different shapes than bottom trawl doors, although doors that can be used with both nets do exist.

Net Structure

When two boats are used (pair trawling), the horizontal spread of the net is provided by the boats, with one or in the case of pelagic trawling two warps attached to each boat. However, single-boat trawling is more common. Here, the horizontal spread of the net is provided by trawl doors (also known as "otter boards"). Trawl doors are available in various sizes and shapes and may be specialized to keep in contact with the sea bottom (bottom trawling) or to remain elevated in the water. In all cases, doors essentially act as wings, using a hydrodynamic shape to provide horizontal spread. As with all wings, the towing vessel must go at a certain speed for the doors to remain standing and functional. This speed varies, but is generally in the range of 2.5–4.0 knots.

Bottom trawling

The vertical opening of a trawl net is created using flotation on the upper edge ("floatline") and weight on the lower edge ("footrope") of the net mouth. The configuration of the footrope varies based on the expected bottom shape. The more uneven the bottom, the more robust the footrope configuration must be to prevent net damage. This is used to catch shrimp, shellfish, cod, scallops and many others. Trawls are funnel-shaped nets that have a closed-off tail where the fish are collected and is open on the top end as the mouth.

Trawl nets can also be modified, such as changing mesh size, to help with marine research of ocean bottoms.

Environmental Effects

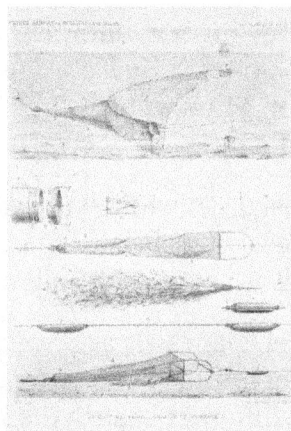

Nets for trawling in surface waters and for trawling in deep water and over the bottom. Note the "tangles" with ensnared marine life

Although trawling today is heavily regulated in some nations, it remains the target of many protests by environmentalists. Environmental concerns related to trawling refer to two areas: the lack of selectivity and the physical damage which the trawl does to the seabed.

Selectivity

Since the practice of trawling started (around the 15th century), there have been concerns over trawling's lack of selectivity. Trawls may be non-selective, sweeping up both marketable and undesirable fish and fish of both legal and illegal size. Any part of the catch which cannot be used is considered by-catch, some of which is killed accidentally by the trawling process. By-catch commonly includes valued species such as dolphins, sea turtles, and sharks, and may also include sublegal or immature individuals of the targeted species.

Many studies have documented large volumes of by-catch that are discarded. For example, researchers conducting a three-year study in the Clarence River found that an estimated 177 tons of by-catch (including 77 different species) were discarded each year.

Size selectivity is controlled by the mesh size of the "cod-end"—the part of the trawl where fish are retained. Fishermen complain that mesh sizes which allow undersized fish to escape also allows some legally–catchable fish to escape as well. There are a number of "fixes", such as tying a rope around the "cod-end" to prevent the mesh from opening fully, which have been developed to work around technical regulation of size selectivity. One problem is when the mesh gets pulled into narrow diamond shapes (rhombuses) instead of squares.

The capture of undesirable species is a recognized problem with all fishing methods and unites environmentalists, who do not want to see fish killed needlessly, and fishermen, who do not want to waste their time sorting marketable fish from their catch. A number of methods to minimize this have been developed for use in trawling. Bycatch reduction grids or square mesh panels of net can be fitted to parts of the trawl, allowing certain species to escape while retaining others.

Studies have suggested that shrimp trawling is responsible for the highest rate of by-catch.

Environmental Damage

Setting a trawl

Trawling is controversial because of its environmental impacts. Because bottom trawling involves towing heavy fishing gear over the seabed, it can cause large-scale destruction on the ocean bottom, including coral shattering, damage to habitats and removal of seaweed. The primary sources of impact are the doors, which can weigh several tonnes and create furrows if dragged along the bottom, and the footrope configuration, which usually remains in contact with the bottom across the entire lower edge of the net. Depending on the configuration, the footrope may turn over large rocks or boulders, possibly dragging them along with the net, disturb or damage sessile organisms or rework and re-suspend bottom sediments. These impacts result in decreases in species diversity and ecological changes towards more opportunistic organisms. The destruction has been likened to clear-cutting in forests.

The primary dispute over trawling concerns the magnitude and duration of these impacts. Opponents argue that they are widespread, intense and long-lasting. Defenders maintain that impact is mostly limited and of low intensity compared to natural events. However, most areas with significant natural sea bottom disturbance events are in relatively shallow water. In mid to deep waters, bottoms trawlers are the only significant area-wide events.

Bottom trawling on soft bottoms also stirs up bottom sediments and loading suspended solids into the water column. One bottom trawler can put more than 10 times the amount of suspended solids pollution per hour into the water column than all the suspended solids pollution from all the sewerage, industrial, river and dredge disposal operations in Southern California combined. These turbidity plumes can be seen on Google Earth in areas where they have high resolution offshore photos. When the turbidity plumes from bottom trawlers are below a thermocline, the surface may not be impacted, but less visible impacts can still occur, such as persistent organic pollutant transfer into the pelagic food chain.

As a result of these processes, a vast array of species are threatened around the world. In particular, trawling can directly kill coral reefs by breaking them up and burying them in sediments. In addition, trawling can kill corals indirectly by wounding coral tissue, leaving the reefs vulnerable to infection. The net effect of fishing practices on global coral reef populations is suggested by many scientists to be alarmingly high. Published research has shown that benthic trawling destroys the cold-water coral *Lophelia pertusa*, an important habitat for many deep-sea organisms.

Midwater (pelagic) trawling is a much "cleaner" method of fishing, in that the catch usually consists of just one species and does not physically damage the sea bottom. However, environmental groups have raised concerns that this fishing practice may be responsible for significant volumes of by-catch, particularly cetaceans (dolphins, porpoises, and whales).

Regulation

In light of the environmental concerns surrounding trawling, many governments have debated policies that would regulate the practice.

Other Uses of the Word "Trawl"

The noun "trawl" has many possibly confusing meanings in commercial fisheries. For example, two or more lobster pots that are fished together may be referred to as a trawl. In some older usages "trawling" meant "long-line fishing"; that usage occurs in Rudyard Kipling's book *Captains Courageous*. (This use is perhaps confused with trolling, where a baited line is trailed behind a boat. Troll also has several meanings.)

The word "trawling" has come to be used in a number of non-fishing contexts, usually meaning indiscriminate collection with the intent of picking out the useful bits. For instance, in law enforcement it may refer to collecting large volumes of telephone call records hoping to find calls made by suspects. It also occurs frequently in reference to research methods, where it means searching through written sources for relevant information.

References

- FAO (2009) Technical Guidelines for Responsible Fisheries, Number 1: Fishing operations, supplement 2 Best practices to reduce incidental catch of seabirds in capture fisheries Rome. ISBN 978-92-5-106423-8.

- Clover, Charles. 2004. The End of the Line: How overfishing is changing the world and what we eat. Ebury Press, London. ISBN 0-09-189780-7

- March, E. J. (1953). Sailing Trawlers: The Story of Deep-Sea Fishing with Long Line and Trawl. Percival Marshal and Company. Reprinted by Charles & David, 1970, Newton Abbot, UK. ISBN 0-7153-4711-X

- FAO (2007) Workshop on standardization of selectivity methods applied to trawling Fisheries Report No. 820. ISBN 978-92-5-005669-2

- Pham, Christopher K; Diogo, Hugo; Menezes, Gui; et al. Deep-water longline fishing has reduced impact on Vulnerable Marine Ecosystems Scientific Reports, via Nature Magazine online. Retrieved 30 December 2015.

- Eigaard B, Thomsen H, Hovgaard H, Nielsen A and Rijnsdorpd AD (2011) "Fishing power increases from technological development in the Faroe Islands longline fishery" Canadian Journal of Fisheries and Aquatic Sciences, 69 (11): 1970–1982. doi:10.1139/f2011-103

Aquaculture: An Integrated Study

This chapter studies aquaculture or aquafarming, the cultivation of aquatic organisms like fish, crustaceans, water plants etc. It delves into the methods and techniques of the commercial industries of fish farming, oyster farming and shrimp farming. It also has a section dedicated to mariculture, the cultivation of marine organisms in open or enclosed saltwater spaces. The aspects elucidated in this chapter are of vital importance, and provide a better understanding of fisheries.

Aquaculture

Aquaculture, also known as aquafarming, is the farming of aquatic organisms such as fish, crustaceans, molluscs and aquatic plants. Aquaculture involves cultivating freshwater and saltwater populations under controlled conditions, and can be contrasted with commercial fishing, which is the harvesting of wild fish. Broadly speaking, the relation of aquaculture to finfish and shellfish fisheries is analogous to the relation of agriculture to hunting and gathering. Mariculture refers to aquaculture practiced in marine environments and in underwater habitats.

Aquaculture installations in southern Chile

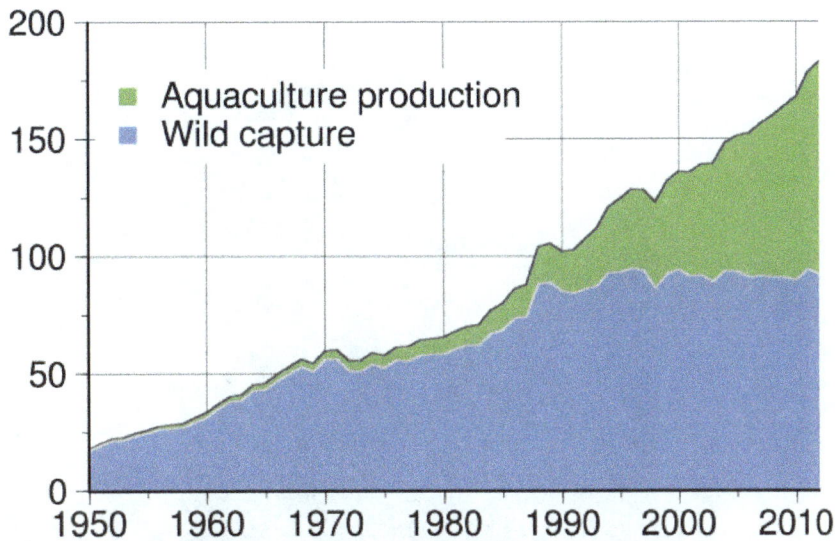

Global harvest of aquatic organisms in million tonnes, 1950–2010, as reported by the FAO

According to the FAO, aquaculture "is understood to mean the farming of aquatic organisms including fish, molluscs, crustaceans and aquatic plants. Farming implies some form of intervention in the rearing process to enhance production, such as regular stocking, feeding, protection from predators, etc. Farming also implies individual or corporate ownership of the stock being cultivated." The reported output from global aquaculture operations would supply one half of the fish and shellfish that is directly consumed by humans; however, there are issues about the reliability of the reported figures. Further, in current aquaculture practice, products from several pounds of wild fish are used to produce one pound of a piscivorous fish like salmon.

Particular kinds of aquaculture include fish farming, shrimp farming, oyster farming, mariculture, algaculture (such as seaweed farming), and the cultivation of ornamental fish. Particular methods include aquaponics and integrated multi-trophic aquaculture, both of which integrate fish farming and plant farming.

History

The indigenous Gunditjmara people in Victoria, Australia, may have raised eels as early as 6000 BC. Evidence indicates they developed about 100 km² (39 sq mi) of volcanic floodplains in the vicinity of Lake Condah into a complex of channels and dams, and used woven traps to capture eels, and preserve them to eat all year round.

Aquaculture was operating in China *circa* 2500 BC. When the waters subsided after river floods, some fish, mainly carp, were trapped in lakes. Early aquaculturists fed their brood using nymphs and silkworm feces, and ate them. A fortunate genetic mutation of carp led to the emergence of goldfish during the Tang dynasty.

Japanese cultivated seaweed by providing bamboo poles and, later, nets and oyster shells to serve as anchoring surfaces for spores.

Romans bred fish in ponds and farmed oysters in coastal lagoons before 100 CE.

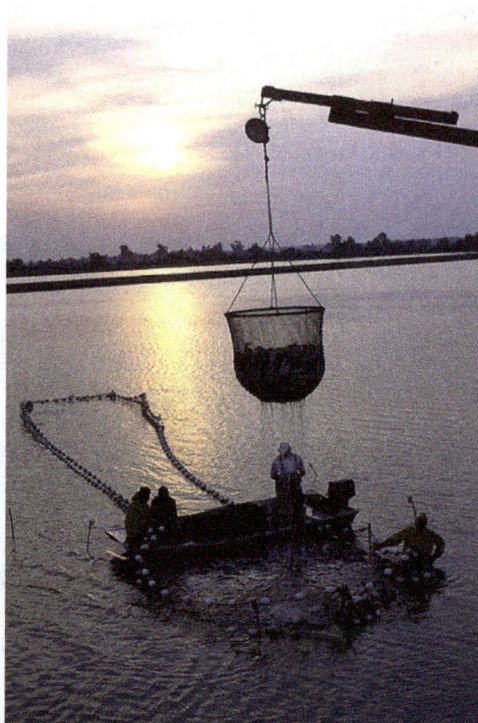

Workers harvest catfish from the Delta Pride Catfish farms in Mississippi

In central Europe, early Christian monasteries adopted Roman aquacultural practices. Aquaculture spread in Europe during the Middle Ages since away from the seacoasts and the big rivers, fish had to be salted so they did not rot. Improvements in transportation during the 19th century made fresh fish easily available and inexpensive, even in inland areas, making aquaculture less popular. The 15th-century fishponds of the Trebon Basin in the Czech Republic are maintained as a UNESCO World Heritage Site.

Hawaiians constructed oceanic fish ponds. A remarkable example is a fish pond dating from at least 1,000 years ago, at Alekoko. Legend says that it was constructed by the mythical Menehune dwarf people.

In first half of 18th century, German Stephan Ludwig Jacobi experimented with external fertilization of brown trouts and salmon. He wrote an article *"Von der künstlichen Erzeugung der Forellen und Lachse"*. By the latter decades of the 18th century, oyster farming had begun in estuaries along the Atlantic Coast of North America.

The word aquaculture appeared in an 1855 newspaper article in reference to the harvesting of ice. It also appeared in descriptions of the terrestrial agricultural practise of subirrigation in the late 19th century before becoming associated primarily with the cultivation of aquatic plant and animal species.

In 1859, Stephen Ainsworth of West Bloomfield, New York, began experiments with brook trout. By 1864, Seth Green had established a commercial fish-hatching operation at Caledonia Springs, near Rochester, New York. By 1866, with the involvement of Dr. W. W. Fletcher of Concord, Massachusetts, artificial fish hatcheries were under way in both Canada and the United States. When the Dildo Island fish hatchery opened in Newfoundland in 1889, it was the largest and most advanced in the world. The word aquaculture was used in descriptions of the hatcheries experiments with cod and lobster in 1890.

By the 1920s, the American Fish Culture Company of Carolina, Rhode Island, founded in the 1870s was one of the leading producers of trout. During the 1940s, they had perfected the method of manipulating the day and night cycle of fish so that they could be artificially spawned year around.

Californians harvested wild kelp and attempted to manage supply around 1900, later labeling it a wartime resource.

21st-century Practice

Harvest stagnation in wild fisheries and overexploitation of popular marine species, combined with a growing demand for high-quality protein, encouraged aquaculturists to domesticate other marine species. At the outset of modern aquaculture, many were optimistic that a "Blue Revolution" could take place in aquaculture, just as the Green Revolution of the 20th century had revolutionized agriculture. Although land animals had long been domesticated, most seafood species were still caught from the wild. Concerned about the impact of growing demand for seafood on the world's oceans, prominent ocean explorer Jacques Cousteau wrote in 1973: "With earth's burgeoning human populations to feed, we must turn to the sea with new understanding and new technology."

About 430 (97%) of the species cultured as of 2007 were domesticated during the 20th and 21st centuries, of which an estimated 106 came in the decade to 2007. Given the long-term importance of agriculture, to date, only 0.08% of known land plant species and 0.0002% of known land animal species have been domesticated, compared with 0.17% of known marine plant species and 0.13% of known marine animal species. Domestication typically involves about a decade of scientific research. Domesticating aquatic species involves fewer risks to humans than do land animals, which took a large toll in human lives. Most major human diseases originated in domesticated animals, including diseases such as smallpox and diphtheria, that like most infectious diseases, move to humans from animals. No human pathogens of comparable virulence have yet emerged from marine species.

Biological control methods to manage parasites are already being used, such as cleaner fish (e.g. lumpsuckers and wrasse) to control sea lice populations in salmon farming. Models are being used to help with spatial planning and siting of fish farms in order to minimize impact.

The decline in wild fish stocks has increased the demand for farmed fish. However, finding alternative sources of protein and oil for fish feed is necessary so the aquaculture industry can grow sustainably; otherwise, it represents a great risk for the over-exploitation of forage fish.

Another recent issue following the banning in 2008 of organotins by the International Maritime Organization is the need to find environmentally friendly, but still effective, compounds with antifouling effects.

Many new natural compounds are discovered every year, but producing them on a large enough scale for commercial purposes is almost impossible.

It is highly probable that future developments in this field will rely on microorganisms, but greater funding and further research is needed to overcome the lack of knowledge in this field.

Species Groups

Global aquaculture production in million tonnes, 1950–2010, as reported by the FAO

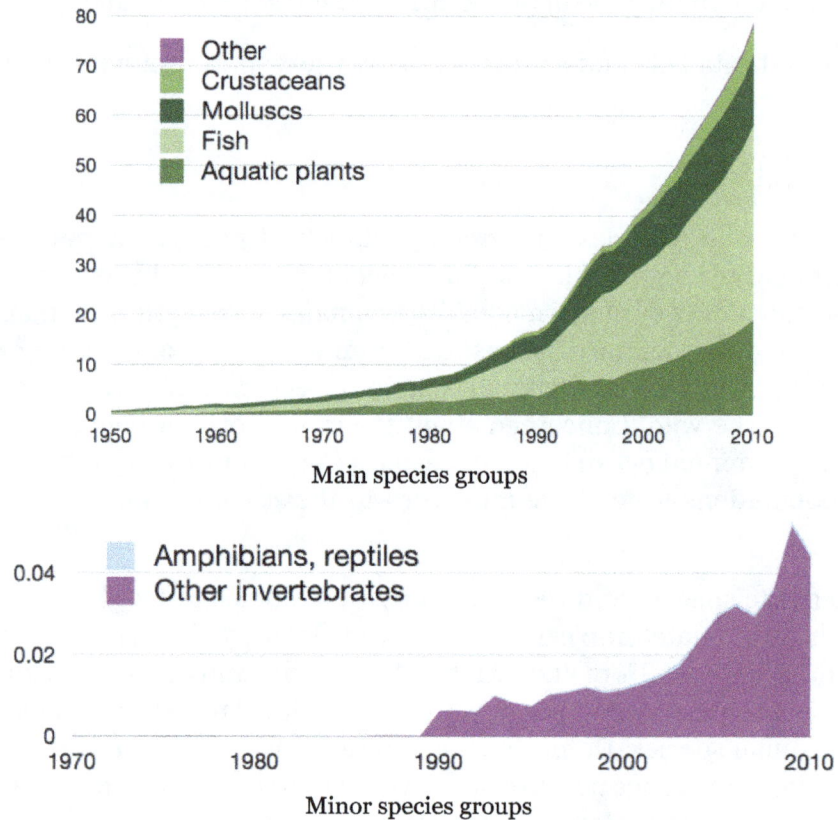

Main species groups

Minor species groups

Aquatic Plants

Cultivating emergent aquatic plants in floating containers

Microalgae, also referred to as phytoplankton, microphytes, or planktonic algae, constitute the majority of cultivated algae. Macroalgae commonly known as seaweed also have many commercial and industrial uses, but due to their size and specific requirements, they are not easily cultivated on a large scale and are most often taken in the wild.

Fish

The farming of fish is the most common form of aquaculture. It involves raising fish commercially in tanks, ponds, or ocean enclosures, usually for food. A facility that releases juvenile fish into the wild for recreational fishing or to supplement a species' natural numbers is generally referred to as a fish hatchery. Worldwide, the most important fish species used in fish farming are, in order, carp, salmon, tilapia, and catfish.

In the Mediterranean, young bluefin tuna are netted at sea and towed slowly towards the shore. They are then interned in offshore pens where they are further grown for the market. In 2009, researchers in Australia managed for the first time to coax southern bluefin tuna to breed in land-locked tanks. Southern bluefin tuna are also caught in the wild and fattened in grow-out sea cages in southern Spencer Gulf, South Australia.

A similar process is used in the salmon-farming section of this industry; juveniles are taken from hatcheries and a variety of methods are used to aid them in their maturation. For example, as stated above, some of the most important fish species in the industry, salmon, can be grown using a cage system. This is done by having netted cages, preferably in open water that has a strong flow, and feeding the salmon a special food mixture that aids their growth. This process allows for year-round growth of the fish, thus a higher harvest during the correct seasons.

Crustaceans

Commercial shrimp farming began in the 1970s, and production grew steeply thereafter. Global production reached more than 1.6 million tonnes in 2003, worth about US$9 billion. About 75% of farmed shrimp is produced in Asia, in particular in China and Thailand. The other 25% is produced mainly in Latin America, where Brazil is the largest producer. Thailand is the largest exporter.

Shrimp farming has changed from its traditional, small-scale form in Southeast Asia into a global industry. Technological advances have led to ever higher densities per unit area, and broodstock is shipped worldwide. Virtually all farmed shrimp are penaeids (i.e., shrimp of the family Penaeidae), and just two species of shrimp, the Pacific white shrimp and the giant tiger prawn, account for about 80% of all farmed shrimp. These industrial monocultures are very susceptible to disease, which has decimated shrimp populations across entire regions. Increasing ecological problems, repeated disease outbreaks, and pressure and criticism from both nongovernmental organizations and consumer countries led to changes in the industry in the late 1990s and generally stronger regulations. In 1999, governments, industry representatives, and environmental organizations initiated a program aimed at developing and promoting more sustainable farming practices through the Seafood Watch program.

Freshwater prawn farming shares many characteristics with, including many problems with, marine shrimp farming. Unique problems are introduced by the developmental lifecycle of the main species, the giant river prawn.

The global annual production of freshwater prawns (excluding crayfish and crabs) in 2003 was about 280,000 tonnes, of which China produced 180,000 tonnes followed by India and Thailand with 35,000 tonnes each. Additionally, China produced about 370,000 tonnes of Chinese river crab.

Molluscs

Abalone farm

Aquacultured shellfish include various oyster, mussel, and clam species. These bivalves are filter and/or deposit feeders, which rely on ambient primary production rather than inputs of fish or other feed. As such, shellfish aquaculture is generally perceived as benign or even beneficial.

Depending on the species and local conditions, bivalve molluscs are either grown on the beach, on longlines, or suspended from rafts and harvested by hand or by dredging.

Abalone farming began in the late 1950s and early 1960s in Japan and China. Since the mid-1990s, this industry has become increasingly successful. Overfishing and poaching have reduced wild populations to the extent that farmed abalone now supplies most abalone meat. Sustainably farmed molluscs can be certified by Seafood Watch and other organizations, including the World Wildlife Fund (WWF). WWF initiated the "Aquaculture Dialogues" in 2004 to develop measurable and performance-based standards for responsibly farmed seafood. In 2009, WWF co-founded the Aquaculture Stewardship Council with the Dutch Sustainable Trade Initiative to manage the global standards and certification programs.

After trials in 2012, a commercial "sea ranch" was set up in Flinders Bay, Western Australia, to raise abalone. The ranch is based on an artificial reef made up of 5000 (As of April 2016) separate concrete units called 'abitats' (abalone habitats). The 900-kg abitats can host 400 abalone each. The reef is seeded with young abalone from an onshore hatchery. The abalone feed on seaweed that has grown naturally on the abitats, with the ecosystem enrichment of the bay also resulting in growing numbers of dhufish, pink snapper, wrasse, and Samson fish, among other species.

Brad Adams, from the company, has emphasised the similarity to wild abalone and the difference from shore-based aquaculture. "We're not aquaculture, we're ranching, because once they're in the water they look after themselves."

Other Groups

Other groups include aquatic reptiles, amphibians, and miscellaneous invertebrates, such as echinoderms and jellyfish. They are separately graphed at the top right of this section, since they do not contribute enough volume to show clearly on the main graph.

Commercially harvested echinoderms include sea cucumbers and sea urchins. In China, sea cucumbers are farmed in artificial ponds as large as 1,000 acres (400 ha).

Around The World

2010

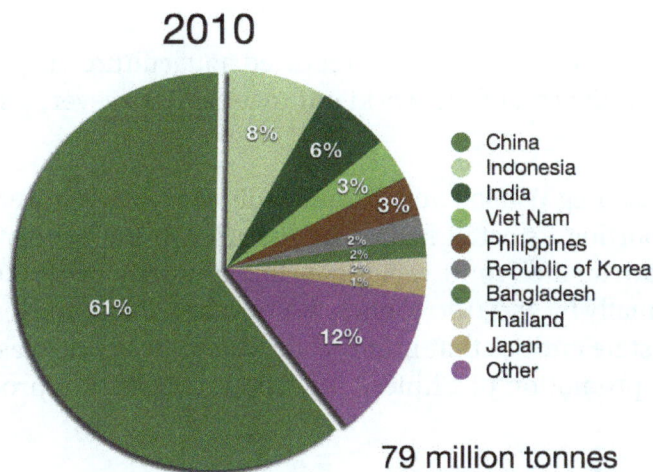

●	China
●	Indonesia
●	India
●	Viet Nam
●	Philippines
●	Republic of Korea
●	Bangladesh
●	Thailand
●	Japan
●	Other

79 million tonnes

Main aquaculture countries in 2010

In 2012, the total world production of fisheries was 158 million tonnes, of which aquaculture contributed 66.6 million tonnes, about 42%. The growth rate of worldwide aquaculture has been sustained and rapid, averaging about 8% per year for over 30 years, while the take from wild fisheries] has been essentially flat for the last decade. The aquaculture market reached $86 billion in 2009.

Aquaculture is an especially important economic activity in China. Between 1980 and 1997, the Chinese Bureau of Fisheries reports, aquaculture harvests grew at an annual rate of 16.7%, jumping from 1.9 million tonnes to nearly 23 million tonnes. In 2005, China accounted for 70% of world production. Aquaculture is also currently one of the fastest-growing areas of food production in the U.S.

About 90% of all U.S. shrimp consumption is farmed and imported. In recent years, salmon aquaculture has become a major export in southern Chile, especially in Puerto Montt, Chile's fastest-growing city.

A United Nations report titled *The State of the World Fisheries and Aquaculture* released in May 2014 maintained fisheries and aquaculture support the livelihoods of some 60 million people in Asia and Africa.

National Laws, Regulations, and Management

Laws governing aquaculture practices vary greatly by country and are often not closely regulated or easily traceable. In the United States, land-based and nearshore aquaculture is regulated at the federal and state levels; however, no national laws govern offshore aquaculture in U.S. exclusive economic zone waters. In June 2011, the Department of Commerce and National Oceanic and Atmospheric Administration released national aquaculture policies to address this issue and "to meet the growing demand for healthy seafood, to create jobs in coastal communities, and restore vital ecosystems." In 2011, Congresswoman Lois Capps introduced the *National Sustainable Offshore Aquaculture Act of 2011* "to establish a regulatory system and research program for sustainable offshore aquaculture in the United States exclusive economic zone"; however, the bill was not enacted into law.

Over-reporting

China overwhelmingly dominates the world in reported aquaculture output, reporting a total output which is double that of the rest of the world put together. However, issues exist with the accuracy of China's returns.

In 2001, fisheries scientists Reg Watson and Daniel Pauly expressed concerns in a letter to *Nature*, that China was over-reporting its catch from wild fisheries in the 1990s. They said that made it appear that the global catch since 1988 was increasing annually by 300,000 tonnes, whereas it was really shrinking annually by 350,000 tonnes. Watson and Pauly suggested this may be related to China policies where state entities that monitor the economy are also asked to increase output. Also, until recently, the promotion of Chinese officials was based on production increases from their own areas.

China disputes this claim. The official Xinhua News Agency quoted Yang Jian, director general of the Agriculture Ministry's Bureau of Fisheries, as saying that China's figures were "basically correct". However, the FAO accepts issues exist with the reliability of China's statistical returns, and currently treats data from China, including the aquaculture data, apart from the rest of the world.

Aquacultural Methods

Mariculture

Mariculture off High Island, Hong Kong

Carp are the dominant fish in aquaculture

The adaptable tilapia is another commonly farmed fish

Mariculture refers to the cultivation of marine organisms in seawater, usually in sheltered coastal waters. The farming of marine fish is an example of mariculture, and so also is the farming of marine crustaceans (such as shrimp), molluscs (such as oysters), and seaweed.

Mariculture may consist of raising the organisms on or in artificial enclosures such as in floating netted enclosures for salmon and on racks for oysters. In the case of enclosed salmon, they are fed by the operators; oysters on racks filter feed on naturally available food. Abalone have been farmed on an artificial reef consuming seaweed which grows naturally on the reef units.

Integrated

Integrated multi-trophic aquaculture (IMTA) is a practice in which the byproducts (wastes) from one species are recycled to become inputs (fertilizers, food) for another. Fed aquaculture (for example, fish, shrimp) is combined with inorganic extractive and organic extractive (for example, shellfish) aquaculture to create balanced systems for environmental sustainability (biomitigation), economic stability (product diversification and risk reduction) and social acceptability (better management practices).

"Multi-trophic" refers to the incorporation of species from different trophic or nutritional levels in the same system. This is one potential distinction from the age-old practice of aquatic polyculture, which could simply be the co-culture of different fish species from the same trophic level. In this case, these organisms may all share the same biological and chemical processes, with few synergistic benefits, which could potentially lead to significant shifts in the ecosystem. Some traditional polyculture systems may, in fact, incorporate a greater diversity of species, occupying several niches, as extensive cultures (low intensity, low management) within the same pond. The term "integrated" refers to the more intensive cultivation of the different species in proximity of each other, connected by nutrient and energy transfer through water.

Ideally, the biological and chemical processes in an IMTA system should balance. This is achieved through the appropriate selection and proportions of different species providing different ecosystem functions. The co-cultured species are typically more than just biofilters; they are harvestable crops of commercial value. A working IMTA system can result in greater total production based on mutual benefits to the co-cultured species and improved ecosystem health, even if the production of individual species is lower than in a monoculture over a short term period.

Sometimes the term "integrated aquaculture" is used to describe the integration of monocultures through water transfer. For all intents and purposes, however, the terms "IMTA" and "integrated aquaculture" differ only in their degree of descriptiveness. Aquaponics, fractionated aquaculture, integrated agriculture-aquaculture systems, integrated peri-urban-aquaculture systems, and integrated fisheries-aquaculture systems are other variations of the IMTA concept.

Netting Materials

Various materials, including nylon, polyester, polypropylene, polyethylene, plastic-coated welded wire, rubber, patented rope products (Spectra, Thorn-D, Dyneema), galvanized steel and copper are used for netting in aquaculture fish enclosures around the world. All of these materials are selected for a variety of reasons, including design feasibility, material strength, cost, and corrosion resistance.

Recently, copper alloys have become important netting materials in aquaculture because they are antimicrobial (i.e., they destroy bacteria, viruses, fungi, algae, and other microbes) and they therefore prevent biofouling (i.e., the undesirable accumulation, adhesion, and growth of microorganisms, plants, algae, tubeworms, barnacles, mollusks, and other organisms). By inhibiting microbial growth, copper alloy aquaculture cages avoid costly net changes that are necessary with other materials. The resistance of organism growth on copper alloy nets also provides a cleaner and healthier environment for farmed fish to grow and thrive.

Issues

Aquaculture can be more environmentally damaging than exploiting wild fisheries on a local area basis but has considerably less impact on the global environment on a per kg of production basis. Local concerns include waste handling, side-effects of antibiotics, competition between farmed and wild animals, and using other fish to feed more marketable carnivorous fish. However, research and commercial feed improvements during the 1990s and 2000s have lessened many of these concerns.

Aquaculture may contribute to propagation of invasive species. As the cases of Nile perch and Janitor fish show, this issue may be damaging to native fauna.

Fish waste is organic and composed of nutrients necessary in all components of aquatic food webs. In-ocean aquaculture often produces much higher than normal fish waste concentrations. The waste collects on the ocean bottom, damaging or eliminating bottom-dwelling life. Waste can also decrease dissolved oxygen levels in the water column, putting further pressure on wild animals.

An alternative model to food being added to the ecosystem, is the installation of artificial reef structures to increase the habitat niches available, without the need to add any more than ambient feed and nutrient. This has been used in the "ranching" of abalone in Western Australia.

Fish Oils

Tilapia from aquaculture has been shown to contain more fat and a much higher ratio of omega-6 to omega-3 oils.

Impacts on wild Fish

Some carnivorous and omnivorous farmed fish species are fed wild forage fish. Although carnivorous farmed fish represented only 13 percent of aquaculture production by weight in 2000, they represented 34 percent of aquaculture production by value.

Farming of carnivorous species like salmon and shrimp leads to a high demand for forage fish to match the nutrition they get in the wild. Fish do not actually produce omega-3 fatty acids, but instead accumulate them from either consuming microalgae that produce these fatty acids, as is the case with forage fish like herring and sardines, or, as is the case with fatty predatory fish, like salmon, by eating prey fish that have accumulated omega-3 fatty acids from microalgae. To satisfy this requirement, more than 50 percent of the world fish oil production is fed to farmed salmon.

Farmed salmon consume more wild fish than they generate as a final product, although the efficiency of production is improving. To produce one pound of farmed salmon, products from several pounds of wild fish are fed to them - this can be described as the "fish-in-fish-out" (FIFO) ratio. In 1995, salmon had a FIFO ratio of 7.5 (meaning 7.5 pounds of wild fish feed were required to produce 1 pound of salmon); by 2006 the ratio had fallen to 4.9. Additionally, a growing share of fish oil and fishmeal come from residues (byproducts of fish processing), rather than dedicated whole fish. In 2012, 34 percent of fish oil and 28 percent of fishmeal came from residues. However, fishmeal and oil from residues instead of whole fish have a different composition with more ash and less protein, which may limit its potential use for aquaculture.

As the salmon farming industry expands, it requires more wild forage fish for feed, at a time when seventy five percent of the worlds monitored fisheries are already near to or have exceeded their maximum sustainable yield. The industrial scale extraction of wild forage fish for salmon farming then impacts the survivability of the wild predator fish who rely on them for food. An important step in reducing the impact of aquaculture on wild fish is shifting carnivorous species to plant-based feeds. Salmon feeds, for example, have gone from containing only fishmeal and oil to containing 40 percent plant protein. The USDA has also experimented with using grain-based feeds for farmed trout. When properly formulated (and often mixed with fishmeal or oil), plant-based feeds can provide proper nutrition and similar growth rates in carnivorous farmed fish.

Another impact aquaculture production can have on wild fish is the risk of fish escaping from coastal pens, where they can interbreed with their wild counterparts, diluting wild genetic stocks. Escaped fish can become invasive, out-competing native species.

Coastal Ecosystems

Aquaculture is becoming a significant threat to coastal ecosystems. About 20 percent of mangrove forests have been destroyed since 1980, partly due to shrimp farming. An extended cost–benefit analysis of the total economic value of shrimp aquaculture built on mangrove ecosystems found that the external costs were much higher than the external benefits. Over four decades, 269,000 hectares (660,000 acres) of Indonesian mangroves have been converted to shrimp farms. Most of these farms are abandoned within a decade because of the toxin build-up and nutrient loss.

Salmon farms are typically sited in pristine coastal ecosystems which they then pollute. A farm with 200,000 salmon discharges more fecal waste than a city of 60,000 people. This waste is discharged directly into the surrounding aquatic environment, untreated, often containing antibiotics and pesticides." There is also an accumulation of heavy metals on the benthos (seafloor) near the salmon farms, particularly copper and zinc.

In 2016, mass fish kill events impacted salmon farmers along Chile's coast and the wider ecology. Increases in aquaculture production and its associated effluent were considered to be possible contributing factors to fish and molluscan mortality.

Genetic Modification

A type of salmon called the AquAdvantage salmon has been genetically modified for faster growth, although it has not been approved for commercial use, due to controversy. The altered salmon incorporates a growth hormone from a Chinook salmon that allows it to reach full size in 16-28 months, instead of the normal 36 months for Atlantic salmon, and while consuming 25 percent less feed. The U.S. Food and Drug Administration reviewed the AquAdvantage salmon in a draft environmental assessment and determined that it "would not have a significant impact (FONSI) on the U.S. environment."

Animal Welfare

As with the farming of terrestrial animals, social attitudes influence the need for humane practices and regulations in farmed marine animals. Under the guidelines advised by the Farm Animal Welfare Council good animal welfare means both fitness and a sense of well being in the animal's physical and mental state. This can be defined by the Five Freedoms:

- Freedom from hunger & thirst

- Freedom from discomfort

- Freedom from pain, disease, or injury

- Freedom to express normal behaviour

- Freedom from fear and distress

However, the controversial issue in aquaculture is whether fish and farmed marine invertebrates are actually sentient, or have the perception and awareness to experience suffering. Although no evidence of this has been found in marine invertebrates, recent studies conclude that fish do have the necessary receptors (nociceptors) to sense noxious stimuli and so are likely to experience states of pain, fear and stress. Consequently, welfare in aquaculture is directed at vertebrates; finfish in particular.

Common Welfare Concerns

Welfare in aquaculture can be impacted by a number of issues such as stocking densities, behavioural interactions, disease and parasitism. A major problem in determining the cause of impaired welfare is that these issues are often all interrelated and influence each other at different times.

Optimal stocking density is often defined by the carrying capacity of the stocked environment and the amount of individual space needed by the fish, which is very species specific. Although behavioural interactions such as shoaling may mean that high stocking densities are beneficial to some species, in many cultured species high stocking densities may be of concern. Crowding can constrain normal swimming behaviour, as well as increase aggressive and competitive behaviours such as cannibalism, feed competition, territoriality and dominance/subordination hierarchies. This potentially increases the risk of tissue damage due to abrasion from fish-to-fish contact or fish-to-cage contact. Fish can suffer reductions in food intake and food conversion efficiency. In addition, high stocking densities can result in water flow being insufficient, creating inadequate oxygen supply and waste product removal. Dissolved oxygen is essential for fish respiration and concentrations below critical levels can induce stress and even lead to asphyxiation. Ammonia, a nitrogen excretion product, is highly toxic to fish at accumulated levels, particularly when oxygen concentrations are low.

Many of these interactions and effects cause stress in the fish, which can be a major factor in facilitating fish disease. For many parasites, infestation depends on the host's degree of mobility, the density of the host population and vulnerability of the host's defence system. Sea lice are the primary parasitic problem for finfish in aquaculture, high numbers causing widespread skin erosion and haemorrhaging, gill congestion, and increased mucus production. There are also a number of prominent viral and bacterial pathogens that can have severe effects on internal organs and nervous systems.

Improving Welfare

The key to improving welfare of marine cultured organisms is to reduce stress to a minimum, as prolonged or repeated stress can cause a range of adverse effects. Attempts to minimise stress can occur throughout the culture process. During grow out it is important to keep stocking densities at appropriate levels specific to each species, as well as separating size classes and grading to reduce aggressive behavioural interactions. Keeping nets and cages clean can assist positive water flow to reduce the risk of water degradation.

Not surprisingly disease and parasitism can have a major effect on fish welfare and it is important for farmers not only to manage infected stock but also to apply disease prevention measures. However, prevention methods, such as vaccination, can also induce stress because of the extra handling and injection. Other methods include adding antibiotics to feed, adding chemicals into water for treatment baths and biological control, such as using cleaner wrasse to remove lice from farmed salmon.

Many steps are involved in transport, including capture, food deprivation to reduce faecal contamination of transport water, transfer to transport vehicle via nets or pumps, plus transport and transfer to the delivery location. During transport water needs to be maintained to a high quality, with regulated temperature, sufficient oxygen and minimal waste products. In some cases anaesthetics may be used in small doses to calm fish before transport.

Aquaculture is sometimes part of an environmental rehabilitation program or as an aid in conserving endangered species.

Prospects

Global wild fisheries are in decline, with valuable habitat such as estuaries in critical condition. The aquaculture or farming of piscivorous fish, like salmon, does not help the problem because they need to eat products from other fish, such as fish meal and fish oil. Studies have shown that salmon farming has major negative impacts on wild salmon, as well as the forage fish that need to be caught to feed them. Fish that are higher on the food chain are less efficient sources of food energy.

Apart from fish and shrimp, some aquaculture undertakings, such as seaweed and filter-feeding bivalve mollusks like oysters, clams, mussels and scallops, are relatively benign and even environmentally restorative. Filter-feeders filter pollutants as well as nutrients from the water, improving water quality. Seaweeds extract nutrients such as inorganic nitrogen and phosphorus directly from the water, and filter-feeding mollusks can extract nutrients as they feed on particulates, such as phytoplankton and detritus.

Some profitable aquaculture cooperatives promote sustainable practices. New methods lessen the risk of biological and chemical pollution through minimizing fish stress, fallowing netpens, and applying Integrated Pest Management. Vaccines are being used more and more to reduce antibiotic use for disease control.

Onshore recirculating aquaculture systems, facilities using polyculture techniques, and properly sited facilities (for example, offshore areas with strong currents) are examples of ways to manage negative environmental effects.

Recirculating aquaculture systems (RAS) recycle water by circulating it through filters to remove fish waste and food and then recirculating it back into the tanks. This saves water and the waste gathered can be used in compost or, in some cases, could even be treated and used on land. While RAS was developed with freshwater fish in mind, scientist associated with the Agricultural Research Service have found a way to rear saltwater fish using RAS in low-salinity waters. Although saltwater fish are raised in off-shore cages or caught with nets in water that typically has a salinity of 35 parts per thousand (ppt), scientists were able to produce healthy pompano, a saltwater fish, in tanks with a salinity of only 5 ppt. Commercializing low-salinity RAS are predicted to have positive environmental and economical effects. Unwanted nutrients from the fish food would not be added to the ocean and the risk of transmitting diseases between wild and farm-raised fish would greatly be reduced. The price of expensive saltwater fish, such as the pompano and combia used in the experiments, would be reduced. However, before any of this can be done researchers must study every aspect of the fish's lifecycle, including the amount of ammonia and nitrate the fish will tolerate in the water, what to feed the fish during each stage of its lifecycle, the stocking rate that will produce the healthiest fish, etc.

Some 16 countries now use geothermal energy for aquaculture, including China, Israel, and the United States. In California, for example, 15 fish farms produce tilapia, bass, and catfish with warm water from underground. This warmer water enables fish to grow all year round and mature more quickly. Collectively these California farms produce 4.5 million kilograms of fish each year.

Fish Farming

Koi farming indoors in Israel

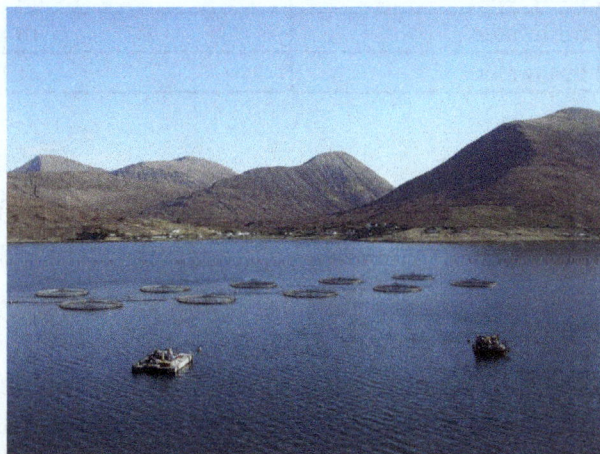

Salmon farming in the sea (mariculture) at Loch Ainort, Isle of Skye

Fish farming or pisciculture involves raising fish commercially in tanks or enclosures, usually for food. It is the principal form of aquaculture, while other methods may fall under mariculture. A facility that releases juvenile fish into the wild for recreational fishing or to supplement a species' natural numbers is generally referred to as a fish hatchery. Worldwide, the most important fish species used in fish farming are carp, tilapia, salmon and catfish.

Demand is increasing for fish and fish protein, which has resulted in widespread overfishing in wild fisheries. China provides 62 percent of the world's farmed fish. As of 2016, more than 50% of seafood was produced by aquaculture.

Farming carnivorous fish, such as salmon, does not always reduce pressure on wild fisheries, since carnivorous farmed fish are usually fed fishmeal and fish oil extracted from wild forage fish. The 2008 global returns for fish farming recorded by the FAO totaled 33.8 million tonnes worth about $US 60 billion.

Major Species

Top 15 cultured fish species by weight in millions of tonnes, according to FAO statistics for 2013			
Species	**Environment**	**Tonnage (millions)**	**Value (USD, billion)**
Grass carp	freshwater	5.23	6.69
Silver carp	freshwater	4.59	6.13
Common carp	freshwater	3.76	5.19
Nile tilapia	freshwater	3.26	5.39
Bighead carp	freshwater	2.90	3.72
Catla (Indian carp)	freshwater	2.76	5.49
Crucian carp	freshwater	2.45	2.67
Atlantic salmon	marine	2.07	10.10
Roho labeo	freshwater	1.57	2.54
Milkfish	freshwater	0.94	1.71
Rainbow trout	freshwater, brackish	0.88	3.80
Wuchang bream	freshwater	0.71	1.16
Black carp	freshwater	0.50	1.15
Northern snakehead	freshwater	0.48	0.59
Amur catfish	freshwater	0.41	0.55

Categories

Aquaculture makes use of local photosynthetical production (extensive) or fish that are fed with external food supply (intensive).

Extensive Aquaculture

Growth is limited by available food, commonly zooplankton feeding on pelagic algae or benthic animals, such as crustaceans and mollusks. Tilapia filter feed directly on phytoplankton, which makes higher production possible. Photosynthetic production can be increased by fertilizing pond water with artificial fertilizer mixtures, such as potash, phosphorus, nitrogen and micro-elements.

Aqua-Boy, a Norwegian live fish carrier used to service the Marine Harvest fish farms on the West coast of Scotland

Another issue is the risk of algal blooms. When temperatures, nutrient supply and available sunlight are optimal for algal growth, algae multiply at an exponential rate, eventually exhausting nutrients and causing a subsequent die-off. The decaying algal biomass will deplete the oxygen in the pond water because it blocks out the sun and pollutes it with organic and inorganic solutes (such as ammonium ions), which can (and frequently do) lead to massive loss of fish.

An alternate option is to use a wetland system such as that of Veta La Palma.

In order to tap all available food sources in the pond, the aquaculturist will choose fish species which occupy different places in the pond ecosystem, e.g., a filter algae feeder such as tilapia, a benthic feeder such as carp or catfish and a zooplankton feeder (various carps) or submerged weeds feeder such as grass carp.

Despite these limitations significant fish farming industries use these methods. In the Czech Republic thousands of natural and semi-natural ponds are harvested each year for trout and carp. The large ponds around Trebon were built from around 1650 and are still in use.

Intensive Aquaculture

Optimal water parameters for cold- and warm-water fish in intensive aquaculture	
Acidity	pH 6-9
Arsenic	<440 µg/L
Alkalinity	>20 mg/L (as $CaCO_3$)
Aluminum	<0.075 mg/L
Ammonia (non-ionized)	<0.02mg/L
Cadmium	<0.0005 mg/L in soft water; <0.005 mg/L in hard water
Calcium	>5 mg/L
Carbon dioxide	<5–10 mg/L
Chloride	>4.0 mg/L
Chlorine	<0.003 mg/L
Copper	<0.0006 mg/L in soft water; <0.03 mg/L in hard water
Gas supersaturation	<100% total gas pressure (103% for salmonid eggs/fry) (102% for lake trout)

Hydrogen sulfide	<0.003 mg/L
Iron	<0.1 mg/L
Lead	<0.02 mg/L
Mercury	<0.0002 mg/L
Nitrate	<1.0 mg/L
Nitrite	<0.1 mg/L
Oxygen	6 mg/L for coldwater fish 4 mg/L for warmwater fish
Selenium	<0.01 mg/L
Total dissolved solids	<200 mg/L
Total suspended solids	<80 NTU over ambient levels
Zinc	<0.005 mg/L

In these kinds of systems fish production per unit of surface can be increased at will, as long as sufficient oxygen, fresh water and food are provided. Because of the requirement of sufficient fresh water, a massive water purification system must be integrated in the fish farm. One way to achieve this is to combine hydroponic horticulture and water treatment. The exception to this rule are cages which are placed in a river or sea, which supplements the fish crop with sufficient oxygenated water. Some environmentalists object to this practice.

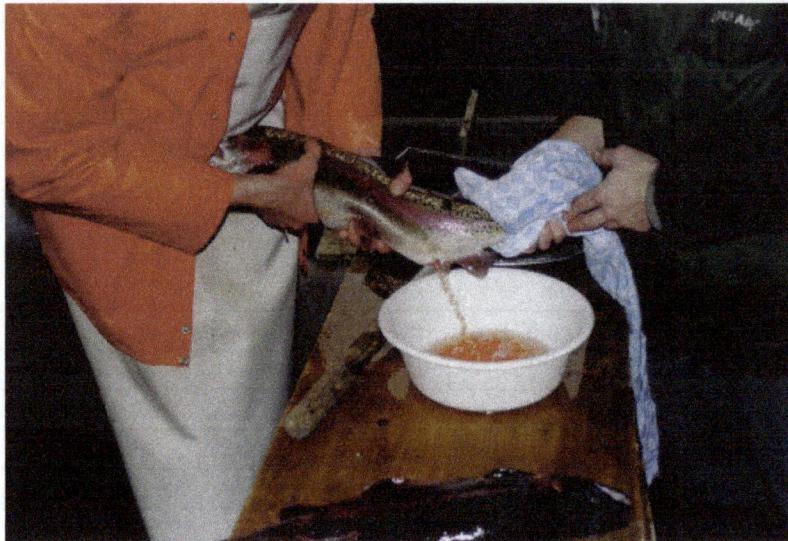

Expressing eggs from a female rainbow trout

The cost of inputs per unit of fish weight is higher than in extensive farming, especially because of the high cost of fish feed, which must contain a much higher level of protein (up to 60%) than cattle food and a balanced amino acid composition as well. However, these higher protein level requirements are a consequence of the higher food conversion efficiency (FCR—kg of feed per kg of animal produced) of aquatic animals. Fish like salmon have an FCR around 1.1 kg of feed per kg of salmon whereas chickens are in the 2.5 kg of feed per kg of chicken range. Fish do not have use energy to keep warm, eliminating a lot of carbohydrates and fats in the diet, required to provide this energy. This however may be offset by the lower land costs and the higher productions which can be obtained due to the high level of input control.

Aeration of the water is essential, as fish need a sufficient oxygen level for growth. This is achieved by bubbling, cascade flow or aqueous oxygen. Catfish, Clarias spp. can breathe atmospheric air and can tolerate much higher levels of pollutants than trout or salmon, which makes aeration and water purification less necessary and makes *Clarias* species especially suited for intensive fish production. In some *Clarias* farms about 10% of the water volume can consist of fish biomass.

The risk of infections by parasites like fish lice, fungi (Saprolegnia spp.), intestinal worms (such as nematodes or trematodes), bacteria (e.g., Yersinia spp., Pseudomonas spp.), and protozoa (such as Dinoflagellates) is similar to animal husbandry, especially at high population densities. However, animal husbandry is a larger and more technologically mature area of human agriculture and better solutions to pathogen problem exist. Intensive aquaculture does have to provide adequate water quality (oxygen, ammonia, nitrite, etc.) levels to minimize stress, which makes the pathogen problem more difficult. This means, intensive aquaculture requires tight monitoring and a high level of expertise of the fish farmer.

Controlling roes manually

Very high intensity recycle aquaculture systems (RAS), where there is control over all the production parameters, are being used for high value species. By recycling the water, very little water is used per unit of production. However, the process does have high capital and operating costs. The higher cost structures mean that RAS is only economical for high value products like broodstock for egg production, fingerlings for net pen aquaculture operations, sturgeon production, research animals and some special niche markets like live fish.

Raising ornamental cold water fish (goldfish or koi), although theoretically much more profitable due to the higher income per weight of fish produced, has never been successfully carried out until very recently. The increased incidences of dangerous viral diseases of koi Carp, together with the high value of the fish has led to initiatives in closed system koi breeding and growing in a number of countries. Today there are a few commercially successful intensive koi growing facilities in the UK, Germany and Israel.

Some producers have adapted their intensive systems in an effort to provide consumers with fish that do not carry dormant forms of viruses and diseases.

In 2016 juvenile Nile tilapia were given a food containing dried *Schizochytrium* in place of fish oil. When compared to a control group raised on regular food, they exhibited higher weight gain and better food-to-growth conversion, plus their flesh was higher in healthy omega-3 fatty acids.

Fish Farms

Within intensive and extensive aquaculture methods, there are numerous specific types of fish farms; each has benefits and applications unique to its design.

Cage System

Giant gourami is often raised in cages in central Thailand

Fish cages are placed in lakes, bayous, ponds, rivers or oceans to contain and protect fish until they can be harvested. The method is also called "off-shore cultivation " when the cages are placed in the sea. They can be constructed of a wide variety of components. Fish are stocked in cages, artificially fed, and harvested when they reach market size. A few advantages of fish farming with cages are that many types of waters can be used (rivers, lakes, filled quarries, etc.), many types of fish can be raised, and fish farming can co-exist with sport fishing and other water uses. Cage farming of fishes in open seas is also gaining popularity. Concerns of disease, poaching, poor water quality, etc., lead some to believe that in general, pond systems are easier to manage and simpler to start. Also, past occurrences of cage-failures leading to escapes, have raised concern regarding the culture of non-native fish species in dam or open-water cages. Even though the cage-industry has made numerous technological advances in cage construction in recent years, storms will always make the concern for escapes valid.

Recently, copper alloys have become important netting materials in aquaculture. Copper alloys are antimicrobial, that is, they destroy bacteria, viruses, fungi, algae, and other microbes. In the marine environment, the antimicrobial/algaecidal properties of copper alloys prevent biofouling, which can briefly be described as the undesirable accumulation, adhesion, and growth of microorganisms, plants, algae, tube worms, barnacles, mollusks, and other organisms.

The resistance of organism growth on copper alloy nets also provides a cleaner and healthier environment for farmed fish to grow and thrive. Traditional netting involves regular and labor-intensive cleaning. In addition to its antifouling benefits, copper netting has strong structural and corrosion-resistant properties in marine environments.

Copper-zinc brass alloys are currently (2011) being deployed in commercial-scale aquaculture operations in Asia, South America and the USA (Hawaii). Extensive research, including demonstrations and trials, are currently being implemented on two other copper alloys: copper-nickel and copper-silicon. Each of these alloy types has an inherent ability to reduce biofouling, cage waste, disease, and the need for antibiotics while simultaneously maintaining water circulation and oxygen requirements. Other types of copper alloys are also being considered for research and development in aquaculture operations.

Irrigation Ditch or Pond Systems

These fish-farming ponds were created as a cooperative project in a rural village.

These use irrigation ditches or farm ponds to raise fish. The basic requirement is to have a ditch or pond that retains water, possibly with an above-ground irrigation system (many irrigation systems use buried pipes with headers.)

Using this method, one can store one's water allotment in ponds or ditches, usually lined with bentonite clay. In small systems the fish are often fed commercial fish food, and their waste products can help fertilize the fields. In larger ponds, the pond grows water plants and algae as fish food. Some of the most successful ponds grow introduced strains of plants, as well as introduced strains of fish.

Control of water quality is crucial. Fertilizing, clarifying and pH control of the water can increase yields substantially, as long as eutrophication is prevented and oxygen levels stay high. Yields can be low if the fish grow ill from electrolyte stress.

Composite Fish Culture

The Composite fish culture system is a technology developed in India by the Indian Council of Agricultural Research in the 1970s. In this system both local and imported fish species, a combination of five or six fish species is used in a single fish pond. These species are selected so that they do not compete for food among them having different types of food habitats. As a result, the food available in all the parts of the pond is used. Fish used in this system include catla and silver carp which are surface feeders, rohu a column feeder and mrigal and common carp which are bottom feeders. Other fish will also feed on the excreta of the common carp and this helps contribute to the efficiency of the system which in optimal conditions will produce 3000–6000 kg of fish per hectare per year.

One problem with such composite fish culture is that many of these fish breed only during monsoon. Even if fish seed is collected from the wild, it can be mixed with that of other species as well. So, a major problem in fish farming is the lack of availability of good-quality seed. To overcome this problem, ways have now been worked out to breed these fish in ponds using hormonal stimulation. This has ensured the supply of pure fish seed in desired quantities.

Integrated Recycling Systems

One of the largest problems with freshwater pisciculture is that it can use a million gallons of water per acre (about 1 m³ of water per m²) each year. Extended water purification systems allow for the reuse (recycling) of local water.

The largest-scale pure fish farms use a system derived (admittedly much refined) from the New Alchemy Institute in the 1970s. Basically, large plastic fish tanks are placed in a greenhouse. A hydroponic bed is placed near, above or between them. When tilapia are raised in the tanks, they are able to eat algae, which naturally grow in the tanks when the tanks are properly fertilized.

The tank water is slowly circulated to the hydroponic beds where the tilapia waste feeds commercial plant crops. Carefully cultured microorganisms in the hydroponic bed convert ammonia to nitrates, and the plants are fertilized by the nitrates and phosphates. Other wastes are strained out by the hydroponic media, which doubles as an aerated pebble-bed filter.

This system, properly tuned, produces more edible protein per unit area than any other. A wide variety of plants can grow well in the hydroponic beds. Most growers concentrate on herbs (e.g. parsley and basil), which command premium prices in small quantities all year long. The most common customers are restaurant wholesalers.

Since the system lives in a greenhouse, it adapts to almost all temperate climates, and may also adapt to tropical climates. The main environmental impact is discharge of water that must be salted to maintain the fishes' electrolyte balance. Current growers use a variety of proprietary tricks to keep fish healthy, reducing their expenses for salt and waste water discharge permits. Some veterinary authorities speculate that ultraviolet ozone disinfectant systems (widely used for ornamental fish) may play a prominent part in keeping the Tilapia healthy with recirculated water.

A number of large, well-capitalized ventures in this area have failed. Managing both the biology and markets is complicated. One future development is the combination of Integrated Recycling systems with Urban Farming as tried in Sweden by the Greenfish initiative.

Classic Fry Farming

This is also called a "Flow through system" Trout and other sport fish are often raised from eggs to fry or fingerlings and then trucked to streams and released. Normally, the fry are raised in long, shallow concrete tanks, fed with fresh stream water. The fry receive commercial fish food in pellets. While not as efficient as the New Alchemists' method, it is also far simpler, and has been used for many years to stock streams with sport fish. European eel (Anguilla anguilla) aquaculturalists procure a limited supply of glass eels, juvenile stages of the European eel which swim north from the Sargasso Sea breeding grounds, for their farms. The European eel is threatened with extinction because of the excessive catch of glass eels by Spanish fishermen and overfishing of adult eels in, e.g., the Dutch IJsselmeer, Netherlands. As per 2005, no one has managed to breed the European eel in captivity.

Issues

The issue of feeds in fish farming has been a controversial one. Many cultured fishes (tilapia, carp, catfish, many others) require no meat or fish products in their diets. Top-level carnivores (most salmon species) depend on fish feed of which a portion is usually derived from wild caught (anchovies, menhaden, etc.). Vegetable-derived proteins have successfully replaced fish meal in feeds for carnivorous fishes, but vegetable-derived oils have not successfully been incorporated into the diets of carnivores.

Secondly, farmed fish are kept in concentrations never seen in the wild (e.g. 50,000 fish in a 2-acre (8,100 m²) area.). However, fish tend also to be animals that aggregate into large schools at high density. Most successful aquaculture species are schooling species, which do not have social problems at high density. Aquaculturists tend to feel that operating a rearing system above its design capacity or above the social density limit of the fish will result in decreased growth rate and increased FCR (food conversion ratio - kg dry feed/kg of fish produced), which will result in increased cost and risk of health problems along with a decrease in profits. Stressing the animals is not desirable, but the concept of and measurement of stress must be viewed from the perspective of the animal using the scientific method.

Sea lice, particularly *Lepeophtheirus salmonis* and various *Caligus* species, including *Caligus clemensi* and *Caligus rogercresseyi*, can cause deadly infestations of both farm-grown and wild salmon. Sea lice are ectoparasites which feed on mucus, blood, and skin, and migrate and latch onto the skin of wild salmon during free-swimming, planktonic *nauplii* and *copepodid* larval stages, which can persist for several days. Large numbers of highly populated, open-net salmon farms can create exceptionally large concentrations of sea lice; when exposed in river estuaries containing large numbers of open-net farms, many young wild salmon are infected, and do not survive as a result. Adult salmon may survive otherwise critical numbers of sea lice, but small, thin-skinned juvenile salmon migrating to sea are highly vulnerable. On the Pacific coast of Canada, the louse-induced mortality of pink salmon in some regions is commonly over 80%.

A 2008 meta-analysis of available data shows that salmon farming reduces the survival of associated wild salmon populations. This relationship has been shown to hold for Atlantic, steelhead, pink, chum, and coho salmon. The decrease in survival or abundance often exceeds 50 percent.

Diseases and parasites are the most commonly cited reasons for such decreases. Some species of sea lice have been noted to target farmed coho and Atlantic salmon. Such parasites have been shown to have an effect on nearby wild fish. One place that has garnered international media attention is British Columbia's Broughton Archipelago. There, juvenile wild salmon must "run a gauntlet" of large fish farms located off-shore near river outlets before making their way to sea. It is alleged that the farms cause such severe sea lice infestations that one study predicted in 2007 a 99% collapse in the wild salmon population by 2011. This claim, however, has been criticized by numerous scientists who question the correlation between increased fish farming and increases in sea lice infestation among wild salmon.

Because of parasite problems, some aquaculture operators frequently use strong antibiotic drugs to keep the fish alive (but many fish still die prematurely at rates of up to 30 percent). In some cases, these drugs have entered the environment. Additionally, the residual presence of these drugs in human food products has become controversial. Use of antibiotics in food production is thought to increase the prevalence of antibiotic resistance in human diseases. At some facilities, the use of antibiotic drugs in aquaculture has decreased considerably due to vaccinations and other techniques. However, most fish farming operations still use antibiotics, many of which escape into the surrounding environment.

The lice and pathogen problems of the 1990s facilitated the development of current treatment methods for sea lice and pathogens. These developments reduced the stress from parasite/pathogen problems. However, being in an ocean environment, the transfer of disease organisms from the wild fish to the aquaculture fish is an ever-present risk.

The very large number of fish kept long-term in a single location contributes to habitat destruction of the nearby areas. The high concentrations of fish produce a significant amount of condensed faeces, often contaminated with drugs, which again affect local waterways. However, if the farm is correctly placed in an area with a strong current, the 'pollutants' are flushed out of the area fairly quickly. Not only does this help with the pollution problem. but water with a stronger current also aids in overall fish growth. Concern remains that resultant bacterial growth strips the water of oxygen, reducing or killing off the local marine life. Once an area has been so contaminated, the fish farms are moved to new, uncontaminated areas. This practice has angered nearby fishermen.

Other potential problems faced by aquaculturists are the obtaining of various permits and water-use rights, profitability, concerns about invasive species and genetic engineering depending on what species are involved, and interaction with the United Nations Convention on the Law of the Sea.

In regards to genetically modified farmed salmon, concern has been raised over their proven reproductive advantage and how it could potentially decimate local fish populations, if released into the wild. Biologist Rick Howard did a controlled laboratory study where wild fish and GMO fish were allowed to breed. The GMO fish crowded out the wild fish in spawning beds, but the offspring were less likely to survive. The colorant used to make pen-raised salmon appear rosy like their wild cousins has been linked with retinal problems in humans.

Labeling

In 2005, Alaska passed legislation requiring that any genetically altered fish sold in the state be labeled. In 2006, a *Consumer Reports* investigation revealed that farm-raised salmon is frequently sold as wild.

In 2008, the US National Organic Standards Board allowed farmed fish to be labeled as organic provided less than 25% of their feed came from wild fish. This decision was criticized by the advocacy group Food & Water Watch as "bending the rules" about organic labeling. In the European Union, fish labeling as to species, method of production and origin, has been required since 2002.

Concerns continue over the labeling of salmon as farmed or wild caught, as well as about the humane treatment of farmed fish. The Marine Stewardship Council has established an Eco label to distinguish between farmed and wild caught salmon, while the RSPCA has established the Freedom Food label to indicate humane treatment of farmed salmon as well as other food products.

Indoor Fish Farming

An alternative to outdoor open ocean cage aquaculture, is through the use of a recirculating aquaculture system (RAS). A RAS is a series of culture tanks and filters where water is continuously recycled and monitored to keep optimal conditions year round. To prevent the deterioration of water quality, the water is treated mechanically through the removal of particulate matter and biologically through the conversion of harmful accumulated chemicals into nontoxic ones.

Other treatments such as UV sterilization, ozonation, and oxygen injection are also used to maintain optimal water quality. Through this system, many of the environmental drawbacks of aquaculture are minimized including escaped fish, water usage, and the introduction of pollutants. The practices also increased feed-use efficiency growth by providing optimum water quality (Timmons et al., 2002; Piedrahita, 2003).

One of the drawbacks to recirculating aquaculture systems is the need for periodic water exchanges. However, the rate of water exchange can be reduced through aquaponics, such as the incorporation of hydroponically grown plants (Corpron and Armstrong, 1983) and denitrification (Klas et al., 2006). Both methods reduce the amount of nitrate in the water, and can potentially eliminate the need for water exchanges, closing the aquaculture system from the environment. The amount of interaction between the aquaculture system and the environment can be measured through the cumulative feed burden (CFB kg/M3), which measures the amount of feed that goes into the RAS relative to the amount of water and waste discharged.

From 2011, a team from the University of Waterloo led by Tahbit Chowdhury and Gordon Graff examined vertical RAS aquaculture designs aimed at producing protein-rich fish species. However, because of its high capital and operating costs, RAS has generally been restricted to practices such as broodstock maturation, larval rearing, fingerling production, research animal production, SPF (specific pathogen free) animal production, and caviar and ornamental fish production. As such, research and design work by Chowdhury and Graff remains difficult to implement. Although the use of RAS for other species is considered by many aquaculturalists to be currently impractical, there has been some limited successful implementation of this with high value product such as barramundi, sturgeon and live tilapia in the US eels and catfish in the Netherlands, trout in Denmark and salmon is planned in Scotland and Canada.

Slaughter Methods

Tanks saturated with carbon dioxide have been used to make fish unconscious. Their gills are then cut with a knife so that the fish bleed out before they are further processed. This is no longer considered a humane method of slaughter. Methods that induce much less physiological stress are electrical or percussive stunning and this has led to the phasing out of the carbon dioxide slaughter method in Europe.

Inhumane Methods

According to T. Håstein of the National Veterinary Institute, "Different methods for slaughter of fish are in place and it is no doubt that many of them may be considered as appalling from an animal welfare point of view." A 2004 report by the EFSA Scientific Panel on Animal Health and Welfare explained: "Many existing commercial killing methods expose fish to substantial suffering over a prolonged period of time. For some species, existing methods, whilst capable of killing fish humanely, are not doing so because operators don't have the knowledge to evaluate them." Following are some of the less humane ways of killing fish.

- *Air Asphyxiation.* This amounts to suffocation in the open air. The process can take upwards of 15 minutes to induce death, although unconsciousness typically sets in sooner.

- *Ice baths / chilling.* Farmed fish are sometimes chilled on ice or submerged in near-freezing water. The purpose is to dampen muscle movements by the fish and to delay the onset of post-death decay. However, it does not necessarily reduce sensibility to pain; indeed, the chilling process has been shown to elevate cortisol. In addition, reduced body temperature extends the time before fish lose consciousness.

- *CO_2 narcosis.*

- *Exsanguination without stunning.* This is a process in which fish are taken up from water, held still, and cut so as to cause bleeding. According to references in Yue, this can leave fish writhing for an average of four minutes, and some catfish still responded to noxious stimuli after more than 15 minutes.

- *Immersion in salt followed by gutting or other processing such as smoking.* This process is applied to eel.

More Humane Methods

Proper stunning renders the fish unconscious immediately and for a sufficient period of time such that the fish is killed in the slaughter process (e.g. through exsanguination) without regaining consciousness.

- *Percussive stunning.* This involves rendering the fish unconscious with a blow on the head.

- *Electric stunning.* This can be humane when a proper current is made to flow through the fish brain for a sufficient period of time. Electric stunning can be applied after the fish has been taken out of the water (dry stunning) or while the fish is still in the water. The latter generally requires a much higher current and may lead to operator safety issues. An advantage could be that in-water stunning allows fish to be rendered unconscious without stressful handling or displacement. However, improper stunning may not induce insensibility long enough to prevent the fish from enduring exsanguination while conscious. It's unknown whether the optimal stunning parameters that researchers have determined in studies are used by the industry in practice.

Shrimp Farming

Shrimp farming is an aquaculture business that exists in either a marine or freshwater environment, producing shrimp or prawns (crustaceans of the groups Caridea or Dendrobranchiata) for human consumption.

Marine

Shrimp grow-out pond on a farm in South Korea

Commercial marine shrimp farming began in the 1970s, and production grew steeply, particularly to match the market demands of the United States, Japan, and Western Europe. The total global production of farmed shrimp reached more than 1.6 million tonnes in 2003, representing a value of nearly US$9 billion. About 75% of farmed shrimp is produced in Asia, particularly in China and Thailand. The other 25% is produced mainly in Latin America, where Brazil, Ecuador, and Mexico are the largest producers. The largest exporting nation is Thailand.

Shrimp farming has changed from traditional, small-scale businesses in Southeast Asia into a global industry. Technological advances have led to growing shrimp at ever higher densities, and broodstock is shipped worldwide. Virtually all farmed shrimp are of the family Penaeidae, and just two species – *Litopenaeus vannamei* (Pacific white shrimp) and *Penaeus monodon* (giant tiger prawn) – account for roughly 80% of all farmed shrimp. These industrial monocultures are very susceptible to diseases, which have caused several regional wipe-outs of farm shrimp populations. Increasing ecological problems, repeated disease outbreaks, and pressure and criticism from both NGOs and consumer countries led to changes in the industry in the late 1990s and generally stronger regulation by governments. In 1999, a program aimed at developing and promoting more sustainable farming practices was initiated, including governmental bodies, industry representatives, and environmental organizations.

Freshwater

A farmer constructing a shrimp farm in Pekalongan, Indonesia

Freshwater prawn farming shares many characteristics with, and many of the same problems as, marine shrimp farming. Unique problems are introduced by the developmental lifecycle of the main species (the giant river prawn, *Macrobrachium rosenbergii*). The global annual production of freshwater prawns in 2010 was about 670,000 tons, of which China produced 615,000 tons (92%).

Animal Welfare

Eyestalk ablation is the removal of one (unilateral) or both (bilateral) eyestalks from a crustacean. It is routinely practiced on female shrimps (or prawns) in almost every marine shrimp maturation or reproduction facility in the world, both research and commercial. The aim of ablation under these circumstances is to stimulate the female shrimp to develop mature ovaries and spawn.

Most captive conditions for shrimp cause inhibitions in females that prevent them from developing mature ovaries. Even in conditions where a given species will develop ovaries and spawn in captivity, use of eyestalk ablation increases total egg production and increases the percentage of females in a given population that participate in reproduction. Once females have been subjected to eyestalk ablation, complete ovarian development often ensues within as little as 3 to 10 days.

Oyster Farming

Oyster farming is an aquaculture (or mariculture) practice in which oysters are raised for human consumption. Oyster farming was practiced by the ancient Romans as early as the 1st century BC on the Italian peninsula and later in Britain for export to Rome. The French oyster industry has relied on aquacultured oysters since the late 18th century.

Harvesting oysters from the pier at Cancale, Brittany, France 2005

History

Oysters farmed in baskets on Prince Edward Island, Canada

Boats used for culturing oysters (circa 1920) in the Gironde estuary, France

Flat bottomed oyster-boat with oyster-bags in Chaillevette, France

Oyster farming was practiced by the ancient Romans as early as the 1st century BC on the Italian peninsula. With the Barbarian invasions the oyster farming in the Mediterranean and the Atlantic came to an end.

In 1852 Monsieur de Bon started to re-seed the oyster beds by collecting the oyster spawn using makeshift catchers. An important step to the modern oyster farming was the oyster farm built by Hyacinthe Boeuf in the Ile de Ré. After obtaining the rights to a part of the coast he built a wall to make a reservoir and to break the strength of the current. Some time later the wall was covered with spat coming spontaneously from the sea which gave 2000 baby oysters per square metre.

Varieties of Farmed Oysters

Commonly farmed food oysters include the Eastern oyster *Crassostrea virginica*, the Pacific oyster *Crassostrea gigas*, Belon oyster *Ostrea edulis*, the Sydney rock oyster *Saccostrea glomerata*, and the Southern mud oyster *Ostrea angasi*.

Cultivation

Oysters naturally grow in estuarine bodies of brackish water. When farmed, the temperature and salinity of the water are controlled (or at least monitored), so as to induce spawning and fertilization, as well as to speed the rate of maturation – which can take several years.

Three methods of cultivation are commonly used. In each case oysters are cultivated to the size of "spat," the point at which they attach themselves to a substrate. The substrate is known as a "cultch" (also spelled "cutch" or "culch"). The loose spat may be allowed to mature further to form "seed" oysters with small shells. In either case (spat or seed stage), they are then set out to mature. The maturation technique is where the cultivation method choice is made.

In one method the spat or seed oysters are distributed over existing oyster beds and left to mature naturally. Such oysters will then be collected using the methods for fishing wild oysters, such as dredging.

In the second method the spat or seed may be put in racks, bags, or cages (or they may be glued in threes to vertical ropes) which are held above the bottom. Oysters cultivated in this manner may be harvested by lifting the bags or racks to the surface and removing mature oysters, or simply retrieving the larger oysters when the enclosure is exposed at low tide. The latter method may avoid losses to some predators, but is more expensive.

In the third method the spat or seed are placed in a cultch within an artificial maturation tank. The maturation tank may be fed with water that has been especially prepared for the purpose of accelerating the growth rate of the oysters. In particular the temperature and salinity of the water may be altered somewhat from nearby ocean water. The carbonate minerals calcite and aragonite in the water may help oysters develop their shells faster and may also be included in the water processing prior to introduction to the tanks. This latter cultivation technique may be the least susceptible to predators and poaching, but is the most expensive to build and to operate. The Pacific oyster *C. gigas* is the species most commonly used with this type of farming.

Oyster culture using tiles as cultch. Taken from The Illustrated London News 1881

Purpose made oyster baskets

Boats

During the nineteenth century in the United States, various shallow draft sailboat designs were developed for oystering in Chesapeake Bay. These included the bugeye, log canoe, pungy, sharpie and skipjack. During the 1880s, a powerboat called the Chesapeake Bay deadrise was also developed.

Since 1977, several boat builders in Brittany have built specialized amphibious vehicles for use in the area's mussel and oyster farming industries. The boats are made of aluminium, are relatively flat-bottomed, and have three, four, or six wheels, depending on the size of the boat. When the tide is out the boats can run on the tidal flats using their wheels. When the tide is in, they use a propeller to move themselves through the water. Oyster farmers in Jersey make use of similar boats. Currently, *Constructions Maritimes du Vivier Amphibie* has a range of models.

Environmental Impact

The farming of oysters and other shellfish is relatively benign or even restorative environmentally, and holds promise for relieving pressure on land-based protein sources. Restoration of oyster populations are encouraged for the ecosystem services they provide, including water quality maintenance, shoreline protection and sediment stabilization, nutrient cycling and sequestration, and habitat for other organisms. A native Olympia oyster restoration project is taking place in Liberty Bay, Washington, and numerous oyster restoration projects are underway in the Chesapeake Bay. In the U.S., Delaware is the only East Coast state without oyster aquaculture, but making aquaculture a state-controlled industry of leasing water by the acre for commercial harvesting of shellfish is being considered. Supporters of Delaware's legislation to allow aquaculture cite revenue, job creation, and nutrient cycling benefits. It is estimated that one acre can produce nearly 750,000 oysters, which could filter between 15 and 40 million gallons of water daily.

Other sources state that a single oyster can filter 24–96 liters a day(1–4 liters per hour). With 750,000 oysters in one acre, 18,000,000-72,000,000 liters of water can be filtered, removing most forms of particulate matter suspended in the water column. The particulate matter oysters remove are sand, clay, silt, detritus, and phytoplankton. These particulates all could possibly contain harmful contamination that originates from anthropogenic sources (the land or directly flowing into the body of water). Instead of becoming ingested by other filter feeders that are then digested by bigger organisms, oysters can sequester these possibly harmful pollutants, and excrete them into the sediment at the bottom of waterways. To remove these contaminants from the sediment, species of seaweed can be added to take up these contaminants in their plant tissues that could be removed and taken to a contained area where the contamination is benign to the surrounding environment.

Predators, Diseases and Pests

Oyster predators include starfish, oyster drill snails, stingrays, Florida stone crabs, birds, such as oystercatchers and gulls, and humans.

Pathogens that can affect either farmed *C. virginica* or *C. gigas* oysters include *Perkinsus marinus* (Dermo) and *Haplosporidium nelsoni* (MSX). However, *C. virginica* are much more susceptible to Dermo or MSX infections than are the *C. gigas* species of oyster. Pathogens of *O. edulis* oysters include *Marteilia refringens* and *Bonamia ostreae*. In the north Atlantic Ocean, oyster crabs may live in an endosymbiotic commensal relationship within a host oyster. Since oyster crabs are considered a food delicacy they may not be removed from young farmed oysters, as they can themselves be harvested for sale.

Polydorid polychaetes are known as pests of cultured oysters.

Mariculture

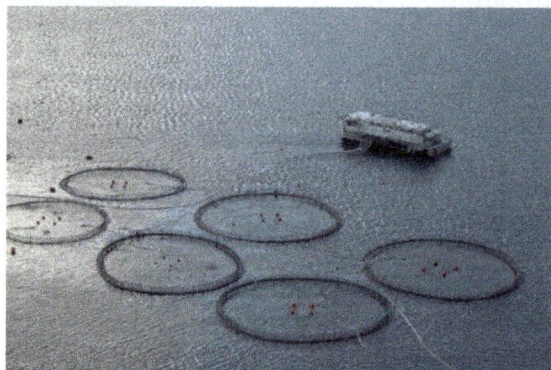

Salmon pens off Vestmanna in the Faroe Islands

Fish cages containing salmon in Loch Ailort, Scotland.

Mariculture is a specialized branch of aquaculture involving the cultivation of marine organisms for food and other products in the open ocean, an enclosed section of the ocean, or in tanks, ponds or raceways which are filled with seawater. An example of the latter is the farming of marine fish, including finfish and shellfish like prawns, or oysters and seaweed in saltwater ponds. Non-food products produced by mariculture include: fish meal, nutrient agar, jewellery (e.g. cultured pearls), and cosmetics.

Methods

Algae

Shellfish

Similar to algae cultivation, shellfish can be farmed in multiple ways: on ropes, in bags or cages, or directly on (or within) the intertidal substrate. Shellfish mariculture does not require feed or fertilizer inputs, nor insecticides or antibiotics, making shellfish aquaculture (or 'mariculture') a self-supporting system. Shellfish can also be used in multi-species cultivation techniques, where shellfish can utilize waste generated by higher trophic level organisms.

Artificial Reefs

After trials in 2012, a commercial "sea ranch" was set up in Flinders Bay, Western Australia to raise abalone. The ranch is based on an artificial reef made up of 5000 (As of April 2016) separate concrete units called *abitats* (abalone habitats). The 900 kilograms (2,000 lb) abitats can host 400 abalone each. The reef is seeded with young abalone from an onshore hatchery.

The abalone feed on seaweed that has grown naturally on the abitats; with the ecosystem enrichment of the bay also resulting in growing numbers of dhufish, pink snapper, wrasse, Samson fish among other species.

Brad Adams, from the company, has emphasised the similarity to wild abalone and the difference from shore based aquaculture. "We're not aquaculture, we're ranching, because once they're in the water they look after themselves."

Open Ocean

Raising marine organisms under controlled conditions in exposed, high-energy ocean environments beyond significant coastal influence, is a relatively new approach to mariculture. Open ocean aquaculture (OOA) uses cages, nets, or long-line arrays that are moored, towed or float freely. Research and commercial open ocean aquaculture facilities are in operation or under development in Panama, Australia, Chile, China, France, Ireland, Italy, Japan, Mexico, and Norway. As of 2004, two commercial open ocean facilities were operating in U.S. waters, raising Threadfin near Hawaii and cobia near Puerto Rico. An operation targeting bigeye tuna recently received final approval. All U.S. commercial facilities are currently sited in waters under state or territorial jurisdiction. The largest deep water open ocean farm in the world is raising cobia 12 km off the northern coast of Panama in highly exposed sites.

Enhanced Stocking

Enchanced Stocking (also known as sea ranching) is a Japanese principle based on operant conditioning and the migratory nature of certain species. The fishermen raise hatchlings in a closely knitted net in a harbor, sounding an underwater horn before each feeding. When the fish are old enough they are freed from the net to mature in the open sea. During spawning season, about 80% of these fish return to their birthplace. The fishermen sound the horn and then net those fish that respond.

Seawater Ponds

In seawater pond mariculture, fish are raised in ponds which receive water from the sea. This has the benefit that the nutrition (e.g. microorganisms) present in the seawater can be used. This is a great advantage over traditional fish farms (e.g. sweet water farms) for which the farmers buy feed (which is expensive). Other advantages are that water purification plants may be planted in the ponds to eliminate the buildup of nitrogen, from fecal and other contamination. Also, the ponds can be left unprotected from natural predators, providing another kind of filtering.

Environmental Effects

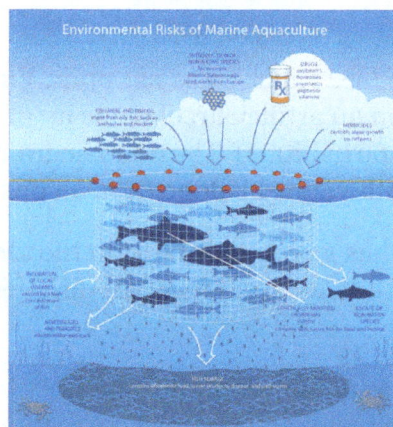

Mariculture has rapidly expanded over the last two decades due to new technology, improvements in formulated feeds, greater biological understanding of farmed species, increased water quality within closed farm systems, greater demand for seafood products, site expansion and government interest. As a consequence, mariculture has been subject to some controversy regarding its social and environmental impacts. Commonly identified environmental impacts from marine farms are:

1. Wastes from cage cultures;

2. Farm escapees and invasives;

3. Genetic pollution and disease and parasite transfer;

4. Habitat modification.

As with most farming practices, the degree of environmental impact depends on the size of the farm, the cultured species, stock density, type of feed, hydrography of the site, and husbandry methods. The adjacent diagram connects these causes and effects.

Wastes from Cage Cultures

Mariculture of finfish can require a significant amount of fishmeal or other high protein food sources. Originally, a lot of fishmeal went to waste due to inefficient feeding regimes and poor digestibility of formulated feeds which resulted in poor feed conversion ratios.

In cage culture, several different methods are used for feeding farmed fish – from simple hand feeding to sophisticated computer-controlled systems with automated food dispensers coupled with *in situ* uptake sensors that detect consumption rates. In coastal fish farms, overfeeding primarily leads to increased disposition of detritus on the seafloor (potentially smothering seafloor dwelling invertebrates and altering the physical environment), while in hatcheries and land-based farms, excess food goes to waste and can potentially impact the surrounding catchment and local coastal environment. This impact is usually highly local, and depends significantly on the settling velocity of waste feed and the current velocity (which varies both spatially and temporally) and depth.

Farm Escapees and Invasives

The impact of escapees from aquaculture operations depends on whether or not there are wild conspecifics or close relatives in the receiving environment, and whether or not the escapee is reproductively capable. Several different mitigation/prevention strategies are currently employed, from the development of infertile triploids to land-based farms which are completely isolated from any marine environment. Escapees can adversely impact local ecosystems through hybridization and loss of genetic diversity in native stocks, increase negative interactions within an ecosystem (such as predation and competition), disease transmission and habitat changes (from trophic cascades and ecosystem shifts to varying sediment regimes and thus turbidity).

The accidental introduction of invasive species is also of concern. Aquaculture is one of the main vectors for invasives following accidental releases of farmed stocks into the wild. One example is the Siberian sturgeon (*Acipenser baerii*) which accidentally escaped from a fish farm into the Gironde Estuary (Southwest France) following a severe storm in December 1999 (5,000 individual fish escaped into the estuary which had never hosted this species before). Molluscan farming is another example whereby species can be introduced to new environments by 'hitchhiking' on farmed molluscs. Also, farmed molluscs themselves can become dominate predators and/or competitors, as well as potentially spread pathogens and parasites.

Genetic Pollution and Disease and Parasite Transfer

One of the primary concerns with mariculture is the potential for disease and parasite transfer. Farmed stocks are often selectively bred to increase disease and parasite resistance, as well as improving growth rates and quality of products. As a consequence, the genetic diversity within reared stocks decreases with every generation – meaning they can potentially reduce the genetic diversity within wild populations if they escape into those wild populations. Such genetic pollution from escaped aquaculture stock can reduce the wild population's ability to adjust to the changing natural environment. Also, maricultured species can harbour diseases and parasites (e.g., lice) which can be introduced to wild populations upon their escape. An example of this is the parasitic sea lice on wild and farmed Atlantic salmon in Canada. Also, non-indigenous species which are farmed may have resistance to, or carry, particular diseases (which they picked up in their native habitats) which could be spread through wild populations if they escape into those wild populations. Such 'new' diseases would be devastating for those wild populations because they would have no immunity to them.

Habitat Modification

With the exception of benthic habitats directly beneath marine farms, most mariculture causes minimal destruction to habitats. However, the destruction of mangrove forests from the farming of shrimps is of concern. Globally, shrimp farming activity is a small contributor to the destruction of mangrove forests; however, locally it can be devastating. Mangrove forests provide rich matrices which support a great deal of biodiversity – predominately juvenile fish and crustaceans. Furthermore, they act as buffering systems whereby they reduce coastal erosion, and improve water quality for in situ animals by processing material and 'filtering' sediments.

Others

In addition, nitrogen and phosphorus compounds from food and waste may lead to blooms of phytoplankton, whose subsequent degradation can drastically reduce oxygen levels. If the algae are toxic, fish are killed and shellfish contaminated.

Sustainability

Mariculture development must be sustained by basic and applied research and development in major fields such as nutrition, genetics, system management, product handling, and socioeconomics. One approach is closed systems that have no direct interaction with the local environment. However, investment and operational cost are currently significantly higher than open cages, limiting them to their current role as hatcheries.

Benefits

Sustainable mariculture promises economic and environmental benefits. Economies of scale imply that ranching can produce fish at lower cost than industrial fishing, leading to better human diets and the gradual elimination of unsustainable fisheries. Maricultured fish are also perceived to be of higher quality than fish raised in ponds or tanks, and offer more diverse choice of species. Consistent supply and quality control has enabled integration in food market channels.

References

- GESAMP (2008) Assessment and communication of environmental risks in coastal aquaculture FAO Reports and Studies No 76. ISBN 978-92-5-105947-0

- Higginbotham, James Arnold (1997-01-01). Piscinae: Artificial Fishponds in Roman Italy. UNC Press Books. ISBN 9780807823293.

- Arnason, Ragnar (2001) Ocean Ranching in Japan In: The Economics of Ocean Ranching: Experiences, Outlook and Theory, FAO, Rome. ISBN 92-5-104631-X.

- Black, K. D. (2001). "Mariculture, Environmental, Economic and Social Impacts of". In Steele, John H.; Thorpe, Steve A.; Turekian, Karl K. Encyclopedia of Ocean Sciences. Academic Press. pp. 1578–1584. doi:10.1006/rwos.2001.0487. ISBN 9780122274305.

- "Information Memorandum, 2013 Ranching of Greenlip Abalone, Flinders Bay – Western Australia" (PDF). Ocean Grown Abalone. Ocean Grown Abalone. Retrieved 23 April 2016.

- Fitzgerald, Bridget (28 August 2014). "First wild abalone farm in Australia built on artificial reef". Australian Broadcasting Corporation Rural. Australian Broadcasting Corporation. Retrieved 23 April 2016. It's the same as the wild core product except we've got the aquaculture advantage which is consistency of supply.

- Murphy, Sean (23 April 2016). "Abalone grown in world-first sea ranch in WA 'as good as wild catch'". Australian Broadcasting Corporation News. Australian Broadcasting Corporation. Retrieved 23 April 2016. So to drive future growth I really believe sea ranching is a great opportunity going forward for some of these coastal communities.

- Brown, Ashton (June 10, 2013). "'Aquaculture' shellfish harvesting bill moves forward". Delaware State News. Retrieved June 11, 2013.

- Lindell, Scott; Miner S; Goudey C; Kite-Powell H; Page S (2012). "Acoustic Conditioning and Ranching of Black Sea Bass Centropristis striata in Massachusetts USA" (PDF). Bull. Fish. Res. Agen. 35: 103–110.

- Data extracted from the FAO Fisheries Global Aquaculture Production Database for freshwater crustaceans. As of October 2012, the most recent data sets are for 2010 and sometimes contain estimates. Accessed October 21, 2012.

- Simon, C.A. 2011. Polydora and Dipolydora (Polychaeta: Spionidae) associated with molluscs on the south coast of South Africa, with descriptions of two new species. African Invertebrates 52 (1): 39-50.

- Yue, Stephanie. "An HSUS Report: The Welfare of Farmed Fish at Slaughter" (PDF). Humane Society of the United States. Retrieved 2011-06-12.

- European Food Safety Authority (2004). "Opinion of the Scientific Panel on Animal Health and Welfare on a request from the Commission related to welfare aspects of the main systems of stunning and killing the main commercial species of animals" (PDF). The EFSA Journal. Retrieved 2011-06-12.

- Håstein, T (2004), "Animal welfare issues relating to aquaculture", Proceedings of the Global Conference on Animal Welfare: an OIE Initiative (PDF), pp. 219–31, retrieved 2011-06-12

- Borgatti, Rachel; Buck, Eugene H. (December 13, 2004). "Open Ocean Aquaculture" (PDF). Congressional Research Service. Retrieved April 10, 2010.

- McAvoy, Audrey (October 24, 2009). "Hawaii regulators approve first US tuna farm". The Associated Press. Retrieved April 9, 2010.

- Schwermer, C. U.; Ferdelman, T. G.; Stief, P.; Gieseke, A.; Rezakhani, N.; Van Rijn, J.; De Beer, D.; Schramm, A. (2010). "Effect of nitrate on sulfur transformations in sulfidogenic sludge of a marine aquaculture biofilter". FEMS Microbiology Ecology. 72 (3): 476–84. doi:10.1111/j.1574-6941.2010.00865.x. PMID 20402774.

Challenges and Threats to Fisheries

The practice of fisheries faces many hindrances and issues that can crop up from overfishing, attack of parasites, unnecessary fish slaughter etc. Conservation and regulation of fish consumption is extremely important to maintain biodiversity of the marine environment. This chapter details these issues with a comprehensive analysis of each topic.

Overfishing

400 tons of jack mackerel caught by a Chilean purse seiner

Overfishing is a form of overexploitation where fish stocks are reduced to below acceptable levels. Overfishing can occur in water bodies of any sizes, such as ponds, rivers, lakes or oceans, and can result in resource depletion, reduced biological growth rates and low biomass levels. Sustained overfishing can lead to critical depensation, where the fish population is no longer able to sustain itself. Some forms of overfishing, for example the overfishing of sharks, has led to the upset of entire marine ecosystems.

The ability of a fishery to recover from overfishing depends on whether the ecosystem's conditions are suitable for the recovery. Dramatic changes in species composition can result in an ecosystem shift, where other equilibrium energy flows involve species compositions different from those that had been present before the depletion of the original fish stock. For example, once trout have been overfished, carp might take over in a way that makes it impossible for the trout to re-establish a breeding population.

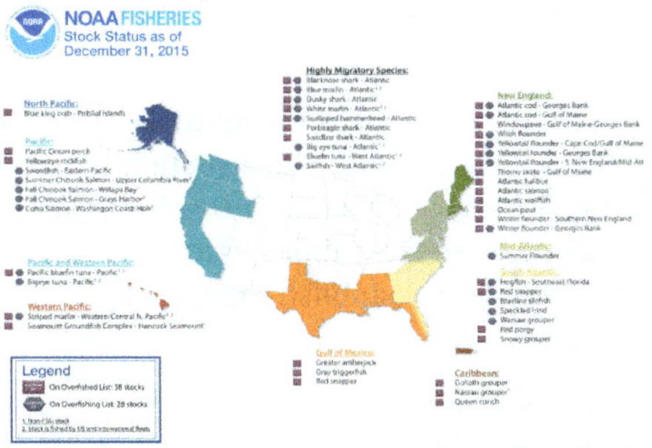

Overfished US stocks, 2015

Overfishing occurs when more fish are caught than the population can replace through natural reproduction. Gathering as many fish as possible may seem like a profitable practice, but overfishing has serious consequences. The results not only affect the balance of life in the oceans, but also the social and economic well-being of the coastal communities who depend on fish for their way of life.

Global Overfishing

Fishing down the food web

Overfishing has significantly affected many fisheries around the world. As much as 85% of the world's fisheries may be over-exploited, depleted, fully exploited or in recovery from exploitation. Significant overfishing has been observed in pre-industrial times. In particular, the overfishing of the western Atlantic Ocean from the earliest days of European colonisation of the Americas has been well documented. Following World War Two, industrial fishing rapidly expanded with rapid increases in worldwide fishing catches. However, many fisheries have either collapsed or degraded to a point where increased catches are no longer possible.

Daniel Pauly, a fisheries scientist known for pioneering work on the human impacts on global fisheries, has commented:

It is almost as though we use our military to fight the animals in the ocean. We are gradually winning this war to exterminate them. And to see this destruction happen, for nothing really – for no reason – that is a bit frustrating. Strangely enough, these effects are all reversible, all the animals that have disappeared would reappear, all the animals that were small would grow, all the relationships that you can't see any more would re-establish themselves, and the system would re-emerge.

Examples of Overfishing

Examples of overfishing exist in areas such as the North Sea, the Grand Banks of Newfoundland and the East China Sea. In these locations, overfishing has not only proved disastrous to fish stocks but also to the fishing communities relying on the harvest. Like other extractive industries such as forestry and hunting, fisheries are susceptible to economic interaction between ownership or stewardship and sustainability, otherwise known as the tragedy of the commons.

- The Peruvian coastal anchovy fisheries crashed in the 1970s after overfishing and an El Niño season largely depleted anchovies from its waters. Anchovies were a major natural resource in Peru; indeed, 1971 alone yielded 10.2 million metric tons of anchovies. However, the following five years saw the Peruvian fleet's catch amount to only about 4 million tons. This was a major loss to Peru's economy.

- The collapse of the cod fishery off Newfoundland, and the 1992 decision by Canada to impose an indefinite moratorium on the Grand Banks, is a dramatic example of the consequences of overfishing.

- The sole fisheries in the Irish Sea, the west English Channel, and other locations have become overfished to the point of virtual collapse, according to the UK government's official Biodiversity Action Plan. The United Kingdom has created elements within this plan to attempt to restore this fishery, but the expanding global human population and the expanding demand for fish has reached a point where demand for food threatens the stability of these fisheries, if not the species' survival.

- Many deep sea fish are at risk, such as orange roughy, Patagonian toothfish, and sablefish. The deep sea is almost completely dark, near freezing and has little food. Deep sea fish grow slowly because of limited food, have slow metabolisms, low reproductive rates, and many don't reach breeding maturity for 30 to 40 years. A fillet of orange roughy at the store is probably at least 50 years old. Most deep sea fish are in international waters, where there are no legal protections. Most of these fish are caught by deep trawlers near seamounts, where they congregate because of food. Flash freezing allows the trawlers to work for days at a time, and modern fishfinders target the fish with ease.

- Blue walleye became extinct in the Great Lakes in the 1980s. Until the middle of the 20th century, it was a commercially valuable fish, with about a half million tonnes being landed during the period from about 1880 to the late 1950s, when the populations collapsed, apparently through a combination of overfishing, anthropogenic eutrophication, and competition with the introduced rainbow smelt.

- The World Wildlife Fund and the Zoological Society of London jointly issued their "Living Blue Planet Report" on 16 September 2015 which states that there was a dramatic fall of 74% in world-wide stocks of the important scombridae fish such as mackerel, tuna and bonitos between 1970 and 2010, and the global overall "population sizes of mammals, birds, reptiles, amphibians and fish fell by half on average in just 40 years."

Examples of Good Fisheries Management

Several countries are now effectively managing their fisheries. Examples include Iceland and New Zealand. The United States has turned many of its fisheries around from being in a highly depleted state.

Consequences

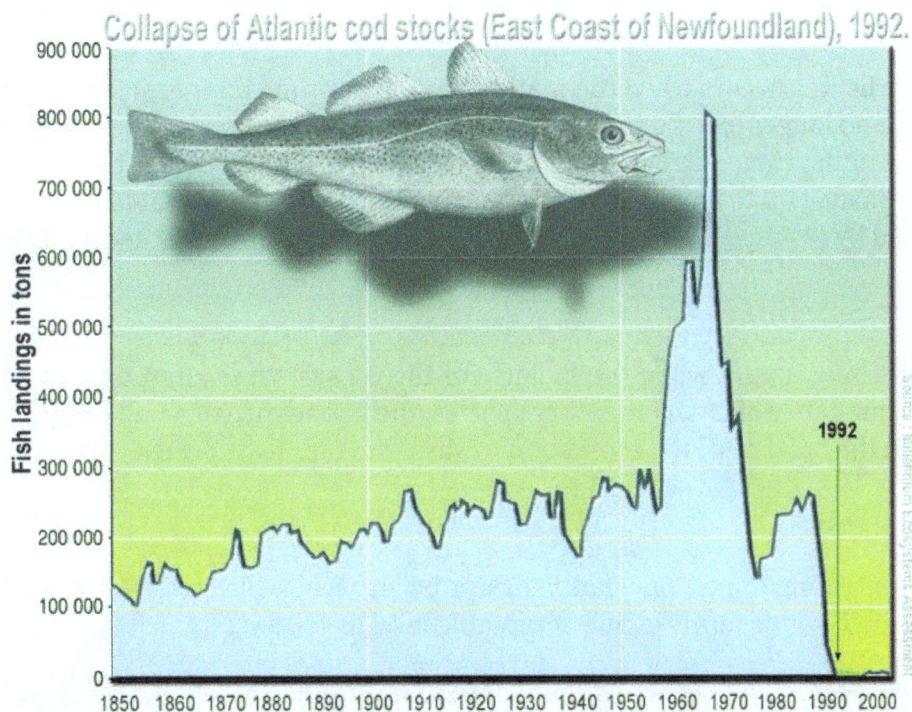

Atlantic cod stocks were severely overfished in the 1970s and 1980s, leading to their abrupt collapse in 1992

According to a 2008 UN report, the world's fishing fleets are losing US$50 billion each year through depleted stocks and poor fisheries management. The report, produced jointly by the World Bank and the UN Food and Agriculture Organization (FAO), asserts that half the world's fishing fleet could be scrapped with no change in catch. In addition, the biomass of global fish stocks have been allowed to run down to the point where it is no longer possible to catch the amount of fish that could be caught. Increased incidence of schistosomiasis in Africa has been linked to declines of fish species that eat the snails carrying the disease-causing parasites. Massive growth of jellyfish populations threaten fish stocks, as they compete with fish for food, eat fish eggs, and poison or swarm fish, and can survive in oxygen depleted environments where fish cannot; they wreak massive havoc on commercial fisheries. Overfishing eliminates a major jellyfish competitor and predator exacerbating the jellyfish population explosion.

Types

There are three recognized types of biological overfishing: growth overfishing, recruit overfishing and ecosystem overfishing.

Growth Overfishing

Growth overfishing occurs when fish are harvested at an average size that is smaller than the size that would produce the maximum yield per recruit. A recruit is an individual that makes it to maturity, or into the limits specified by a fishery, which are usually size or age. This makes the total yield less than it would be if the fish were allowed to grow to an appropriate size. It can be countered by reducing fishing mortality to lower levels and increasing the average size of harvested fish to a size that will allow maximum yield per recruit.

Recruitment Overfishing

Recruitment overfishing occurs when the mature adult (spawning biomass) population is depleted to a level where it no longer has the reproductive capacity to replenish itself—there are not enough adults to produce offspring. Increasing the spawning stock biomass to a target level is the approach taken by managers to restore an overfished population to sustainable levels. This is generally accomplished by placing moratoriums, quotas and minimum size limits on a fish population.

Ecosystem Overfishing

Ecosystem overfishing occurs when the balance of the ecosystem is altered by overfishing. With declines in the abundance of large predatory species, the abundance of small forage type increases causing a shift in the balance of the ecosystem towards smaller fish species.

Acceptable Levels

The notion of overfishing hinges on what is meant by an acceptable level of fishing. More precise biological and bioeconomic terms define acceptable level as follows:

- Biological overfishing occurs when fishing mortality has reached a level where the stock biomass has negative marginal growth (reduced rate of biomass growth), as indicated by the red area in the figure. (Fish are being taken out of the water so quickly that the replenishment of stock by breeding slows down. If the replenishment continues to diminish for long enough, replenishment will go into reverse and the population will decrease.)

- Economic or bioeconomic overfishing additionally considers the cost of fishing when determining acceptable catches. Under this framework, a fishery is considered to be overfished when catches exceed maximum economic yield where resource rent is at its maximum. Fish are being removed from the fishery so quickly that the profitability of the fishery is sub-optimal. A more dynamic definition of economic overfishing also considers the present value of the fishery using a relevant discount rate to maximise the flow of resource rent over all future catches.

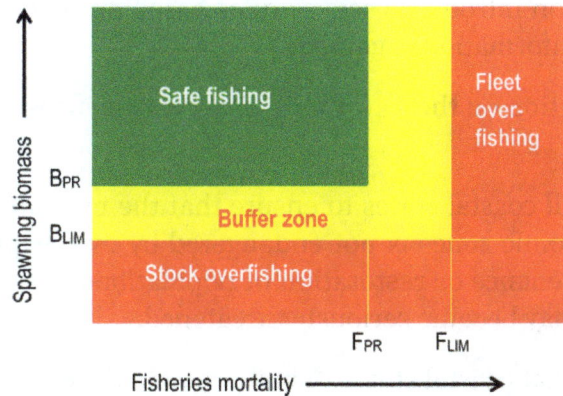

The Traffic Light colour convention, showing the concept of Harvest Control Rule (HCR), specifying when a rebuilding plan is mandatory in terms of precautionary and limit reference points for spawning biomass and fishing mortality rate.

Harvest Control Rule

A model proposed in 2010 for predicting acceptable levels of fishing is the Harvest Control Rule (HCR), which is a set of tools and protocols with which management has some direct control of harvest rates and strategies in relation to predicting stock status, and long-term maximum sustainable yields. Constant catch and constant fishing mortality are two types of simple harvest control rules.

Input and Output Orientations

Fishing capacity can also be defined using an input or output orientation.

- An input-oriented fishing capacity is defined as the maximum available capital stock in a fishery that is fully utilized at the maximum technical efficiency in a given time period, given resource and market conditions.

- An output-oriented fishing capacity is defined as the maximum catch a vessel (fleet) can produce if inputs are fully utilized given the biomass, the fixed inputs, the age structure of the fish stock, and the present stage of technology.

Technical efficiency of each vessel of the fleet is assumed necessary to attain this maximum catch. The degree of capacity utilization results from the comparison of the actual level of output (input) and the capacity output (input) of a vessel or a fleet.

Mitigation

With present and forecast world population levels it is not possible to solve the over fishing issue; however, there are mitigation measures that can save selected fisheries and forestall the collapse of others.

In order to meet the problems of overfishing, a precautionary approach and Harvest Control Rule (HCR) management principles have been introduced in the main fisheries around the world. The

Traffic Light color convention introduces sets of rules based on predefined critical values, which could be adjusted as more information is gained.

The United Nations Convention on the Law of the Sea treaty deals with aspects of over fishing in articles 61, 62, and 65.

- Article 61 requires all coastal states to ensure that the maintenance of living resources in their exclusive economic zones is not endangered by over-exploitation. The same article addresses the maintenance or restoration of populations of species above levels at which their reproduction may become seriously threatened.

- Article 62 provides that coastal states: "shall promote the objective of optimum utilization of the living resources in the exclusive economic zone without prejudice to Article 61"

- Article 65 provides generally for the rights of, inter alia, coastal states to prohibit, limit, or regulate the exploitation of marine mammals.

According to some observers, overfishing can be viewed as an example of the tragedy of the commons; appropriate solutions would therefore promote property rights through, for instance, privatization and fish farming. Daniel K. Benjamin, in *Fisheries are Classic Example of the "Tragedy of the Commons"*, cites research by Grafton, Squires and Fox to support the idea that privatization can solve the overfishing problem:

> *According to recent research on the British Columbia halibut fishery, where the commons has been at least partly privatized, substantial ecological and economic benefits have resulted. There is less damage to fish stocks, the fishing is safer, and fewer resources are needed to achieve a given harvest.*

Another possible solution, at least for some areas, is quotas, so fishermen can only legally take a certain amount of fish. A more radical possibility is declaring certain areas of the sea "no-go zones" and make fishing there strictly illegal, so the fish in that area have time to recover and repopulate.

Controlling consumer behavior and demand is a key in mitigating action. Worldwide, a number of initiatives emerged to provide consumers with information regarding the conservation status of the seafood available to them. The Guide to Good Fish Guides lists a number of these.

Government Regulation

Many regulatory measures are available for controlling overfishing. These measures include fishing quotas, bag limits, licensing, closed seasons, size limits and the creation of marine reserves and other marine protected areas.

A model of the interaction between fish and fishers showed that when an area is closed to fishers, but there are no catch regulations such as individual transferable quotas, fish catches are temporarily increased but overall fish biomass is reduced, resulting in the opposite outcome from the one desired for fisheries. Thus, a displacement of the fleet from one locality to another will generally have little effect if the same quota is taken. As a result, management measures such as temporary closures or establishing a marine protected area of fishing areas are ineffective when not combined with individual fishing quotas. An inherent problem with quotas is that fish populations vary from year to year. A study has found that fish populations rise dramatically after stormy years due to more nutrients reaching the surface and therefore greater primary production. To fish sustainably, quotas need to be changed each year to account for fish population.

Individual transferable quotas (ITQs) are fishery rationalization instruments defined under the Magnuson-Stevens Fishery Conservation and Management Act as limited access permits to harvest quantities of fish. Fisheries scientists decide the optimal amount of fish (total allowable catch) to be harvested in a certain fishery. The decision considers carrying capacity, regeneration rates and future values. Under ITQs, members of a fishery are granted rights to a percentage of the total allowable catch that can be harvested each year. These quotas can be fished, bought, sold, or leased allowing for the least cost vessels to be used. ITQs are used in New Zealand, Australia, Iceland, Canada, and the United States. Only three ITQ programs have been implemented in the United States due to a moratorium supported by Ted Stevens.

In 2008, a large-scale study of fisheries that used ITQs compared to ones that didn't provided strong evidence that ITQs can help to prevent collapses and restore fisheries that appear to be in decline.

China bans fishing in the South China Sea for a period each year.

Removal of Subsidies

Several scientists have called for an end to subsidies paid to deep sea fisheries. In international waters beyond the 200 nautical mile exclusive economic zones of coastal countries, many fisheries are unregulated, and fishing fleets plunder the depths with state-of-the-art technology. In a few hours, massive nets weighing up to 15 tons, dragged along the bottom by deep-water trawlers, can destroy deep-sea corals and sponge beds that have taken centuries or millennia to grow. The trawlers can target orange roughy, grenadiers, or sharks. These fish are usually long-lived and late maturing, and their populations take decades, even centuries to recover.

Fisheries scientist Daniel Pauly and economist Ussif Rashid Sumaila have examined subsidies paid to bottom trawl fleets around the world. They found that US$152 million per year are paid to deep-sea fisheries. Without these subsidies, global deep-sea fisheries would operate at a loss of $50 million a year. A great deal of the subsidies paid to deep-sea trawlers is to subsidize the large amount of fuel required to travel beyond the 200-mile limit and drag weighted nets.

"There is surely a better way for governments to spend money than by paying subsidies to a fleet that burns 1.1 billion litres of fuel annually to maintain paltry catches of old growth fish from highly vulnerable stocks, while destroying their habitat in the process" – *Pauly*.

"Eliminating global subsidies would render these fleets economically unviable and would relieve tremendous pressure on over-fishing and vulnerable deep-sea ecosystems" – *Sumaila*.

Minimizing Fishing Impact

Fishing techniques may be altered to minimize bycatch and reduce impacts on marine habitats. These techniques include using varied gear types depending on target species and habitat type. For example, a net with larger holes will allow undersized fish to avoid capture. A turtle excluder device

(TED) allows sea turtles and other megafauna to escape from shrimp trawls. Avoiding fishing in spawning grounds may allow fish stocks to rebuild by giving adults a chance to reproduce.

Aquaculture

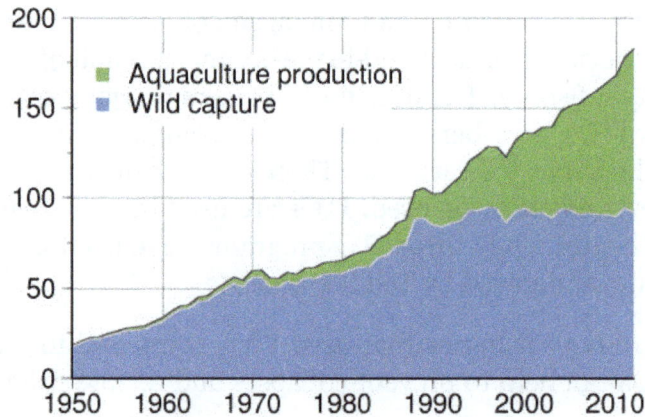

Global harvest of aquatic organisms in million tonnes, 1950–2010, as reported by the FAO.

Aquaculture involves the farming of fish in captivity. This approach effectively privatizes fish stocks and creates incentives for farmers to conserve their stocks. It also reduces environmental impact. However, farming carnivorous fish, such as salmon, does not always reduce pressure on wild fisheries, since carnivorous farmed fish are usually fed fishmeal and fish oil extracted from wild forage fish.

Aquaculture played a minor role in the harvesting of marine organisms until the 1970s. Growth in aquaculture increased rapidly in 1990s when the rate of wild capture plateaued. Aquaculture now provides approximately half of all harvested aquatic organisms. Aquaculture production rates continue to grow while wild harvest remains steady.

Fish farming can enclose the entire breeding cycle of the fish, with fish being bred in captivity. Some fish prove difficult to breed in captivity and can be caught in the wild as juveniles and brought into captivity to increase their weight. With scientific progress more species are being made to breed in captivity. This was the case with southern bluefin tuna, which were first bred in captivity in 2009.

Consumer Awareness

Sustainable seafood is a movement that has gained momentum as more people become aware of overfishing and environmentally destructive fishing methods. Sustainable seafood is seafood from either fished or farmed sources that can maintain or increase production in the future without jeopardizing the ecosystems from which it was acquired. In general, slow-growing fish that reproduce late in life, such as orange roughy, are vulnerable to overfishing. Seafood species that grow quickly and breed young, such as anchovies and sardines, are much more resistant to overfishing. Several organizations, including the Marine Stewardship Council (MSC), and Friend of the Sea, certify seafood fisheries as sustainable.

The Marine Stewardship Council has developed an environmental standard for sustainable and well-managed fisheries. Environmentally responsible fisheries management and practices are rewarded with the use of its blue product ecolabel. Consumers concerned about overfishing and its consequences are increasingly able to choose seafood products that have been independently assessed against the MSC's environmental standard. This enables consumers to play a part in reversing the decline of fish stocks. As of February 2012, over 100 fisheries around the world have been independently assessed and certified as meeting the MSC standard. Their where to buy page lists the currently available certified seafood. As of February 2012 over 13,000 MSC-labelled products are available in 74 countries around the world. Fish & Kids is an MSC project to teach schoolchildren about marine environmental issues, including overfishing.

The Monterey Bay Aquarium's Seafood Watch Program, although not an official certifying body like the MSC, also provides guidance on the sustainability of certain fish species. Some seafood restaurants have begun to offer more sustainable seafood options. The Seafood Choices Alliance is an organization whose members include chefs that serve sustainable seafood at their establishments. In the US, the Sustainable Fisheries Act defines sustainable practices through national standards. Although there is no official certifying body like the MSC, the National Oceanic and Atmospheric Administration has created FishWatch to help guide concerned consumers to sustainable seafood choices.

Awareness in the Fishing Community

Creating awareness of overfishing and effective measures can be effective in fisheries management. Improving awareness of regulations can improve compliance. Also, creating public awareness of environmental impacts of fishing can lead to fishermen voluntarily engaging in practices such as catch and release.

Barriers to Effective Fishery Management

The fishing industry has a strong financial incentive to oppose some measures aimed at improving the sustainability of fish stocks. Recreational fisherman also have an interest in maintaining access to fish stocks. This leads to extensive lobbying that can block or water down government policies intended to prevent overfishing.

Outside of countries' exclusive economic zones, fishing is difficult to control. Large oceangoing fishing boats are free to exploit fish stocks at will.

In waters that are the subject of territorial disputes, countries may actively encourage overfishing. A notable example is the cod wars where Britain used its navy to protect its trawlers fishing in Iceland's exclusive economic zone. Fish are highly transitory. Many species will freely move through different jurisdictions. The conservation efforts of one country can then be exploited by another.

While governments can create regulations to control people's behaviours this can be undermined by illegal fishing activity. Estimates of the size of the illegal catch range from 11 to 26 million tonnes, which represents 14-33% of the world's reported catch. Illegal fishing can take many forms. In some developing countries, large numbers of poor people are dependent on fishing. It can prove difficult to regulate this kind of overfishing, especially for weak governments. Even in regulated environments, illegal fishing may occur. While industrial fishing is often effectively controlled, smaller scale and recreational fishermen can often break regulations such as bag limits and seasonal closures. Some illegal fishing takes place on an industrial scale with financed commercial operations.

The fishing capacity problem is not only related to the conservation of fish stocks but also to the sustainability of fishing activity. Causes of the fishing problem can be found in the property rights regime of fishing resources. Overexploitation and rent dissipation of fishermen arise in open-access fisheries as was shown in Gordon.

In open-access resources like fish stocks, in the absence of a system like individual transferable quotas, the impossibility of excluding others provokes the fishermen who want to increase catch to do so effectively by taking someone else' share, intensifying competition. This tragedy of the commons provokes a capitalization process that leads them to increase their costs until they are equal to their revenue, dissipating their rent completely.

Resistance From Fishermen

There is always disagreement between fishermen and government scientists... Imagine an over-fished area of the sea in the shape of a hockey field with nets at either end. The few fish left therein would gather around the goals because fish like structured habitats. Scientists would survey the entire field, make lots of unsuccessful hauls, and conclude that it contains few fish. The fishermen would make a beeline to the goals, catch the fish around them, and say the scientists do not know what they are talking about. The subjective impression the fishermen get is always that there's lots of fish - because they only go to places that still have them... fisheries scientists survey and compare entire areas, not only the productive fishing spots. – *Fisheries scientist Daniel Pauly*

Fish Slaughter

Fish slaughter is the process of killing fish, typically after harvesting at sea or from fish farms. At least one trillion fish are slaughtered each year for human consumption. Some relatively humane slaughter methods have been developed, including percussive and electric stunning. However, most fish harvesting continues to use methods like suffocation in air, carbon-dioxide stunning, or ice chilling that may not optimise fish welfare in some instances.

Numbers

According to the Food and Agriculture Organization (FAO), a total of 156.2 million tons of fish, crustaceans, molluscs, and other aquatic animals were captured in 2011. This is a sum of 93.5 million tons of wild animals and 62.7 million tons of farmed animals. 56.8% of this total was freshwater fish, 6.4% diadromous fish, and 3.2% marine fish, with the remainder being molluscs, crustaceans, and miscellaneous.

The number of individual wild fish killed each year is estimated as 0.97-2.74 trillion (based on FAO tonnage statistics combined with estimated mean weights of fish species). The FAO numbers do not include illegal, unreported and unregulated fishing, nor discarded fish. If these are included and over-reporting by China subtracted, the totals increase by about 16.6%

to 33.3%. A similar estimate for the number of farmed fish slaughtered each year is 0.037 to 0.120 trillion.

Mid-sized trout farms in the UK may process more than 10,000 fish per hour. They are often operated by only a few people, and it may be necessary to kill trout on short notice or even at night.

Welfare Indicators

Research on fish suffering during slaughter relies on measures to indicate when fish are stressed. Some indicators used by welfare studies include

- Behavior

    o Swimming, gill movement, eye movement in response to body re-orientation, reaction when inverted, etc.

- Electrical measures

    o EEG, ECG, evoked responses, etc.

    o These are quite accurate but also require high levels of expertise.

- Hematic measures

    o Cortisol, plasma glucose, plasma lactate, hematocrit, etc.

- Tissue measures

    o Indicators of stress in the muscle tissue, like lactic acid, pH, and the catabolites of adenosine triphosphate (ATP).

    o These indicators typically also correlate with lower-quality meat.

Following electric stunning, as fish gradually resume consciousness, they begin to make rhythmic gill-cover movements. Based on EEG correlations, it is believed that stunned fish remain insensible until they have resumed rhythmic gill patterns. This can be used as a convenient assessment tool for the effectiveness of electric stunning.

Inhumane Methods

In 2004, the European Food Safety Authority observed that "Many existing commercial killing methods expose fish to substantial suffering over a prolonged period of time."

The *Aquatic Animal Health Code* of the World Organisation for Animal Health considers the following slaughter methods inhumane.

Air Asphyxiation

This is the oldest slaughter method for fish and is considered inhumane because it can take the fish over an hour to die. One Dutch study found that it took 55–250 minutes for various species of fish to become insensible during asphyxiation. Fish that evolved for low-oxygen environments take longer to die. At higher temperatures, fish lose consciousness more quickly.

Meat quality and shelf-life are also diminished when this method is used.

Ice Bath

Also called *live chilling*, this method involves putting fish in baths of ice water, where they chill and eventually die of anoxia. Because chilling slows metabolic rate and oxygen needs, it may pro-long the duration until death in some instances, with some cold adapted species taking more than an hour to die. On these grounds, the Farm Animal Welfare Council's 1996 report on farmed-fish welfare stated: "The cooling of live trout on ice after they have been removed from water should be prohibited." In contrast, later research suggested that for warm Mediterranean species such as sea bream and sea bass, the method might at least be preferable to air asphyxiation, with fish showing lower levels of stress indicators. More recent research has shown that ice water is faster and less stressful than anaesthetics for killing tropical ornamental fishes like zebrafish.

Co$_2$ Narcosis

Most often applied for salmon and trout, CO_2 narcosis involves filling the fish water with CO_2 to produce acidic pH, which injures the brain. The procedure is apparently stressful, as evidenced by fish swimming vigorously and trying to escape from the tank. CO_2 immobilizes the fish within 2–4 minutes, but the fish remain conscious until subsequent stunning or killing.

Salt or Ammonia Baths

Exsanguination Without Stunning

Exsanguination is the process whereby an animal is cut so that it bleeds to death. Fish are cut in highly vascular body regions, and the process is stressful unless the animals are unconscious. If not stunned, according to behavioral and neural criteria, fish may remain conscious for 15 minutes or more between the time when major blood vessels have been cut and when they lose consciousness. Eel brains may continue to process information for 13–30 minutes after being decapitated, and some fish may remain sensible for 20–40 minutes after evisceration.

Potentially More Humane Methods

Percussive Stunning

Also known as *knocking*, percussive stunning involves hitting the fish's head with a wooden or plastic club, called a *priest*. One or two appropriate blows can disrupt the brain sufficiently to render the fish insensible and potentially even kill it directly. However, applying this method correctly requires training and effort. Percussive stunning must be applied one fish at a time and so is typically only used for large fish, such as salmon and trout. If the operator is skilled, percussive stunning can be among the most humane methods and can also yield high meat quality. One comparison of slaughter methods found that percussive stunning had the best welfare performance as measured by low hematocrit, low plasma glucose, low lactate, and high muscle energy charge.

For some fish species, there are automated percussive stunning tools, such as a pneumatic club for salmon. However, building an automated machine to process, orient, percussively stun, and bleed bulk quantities of small fish would be difficult.

Pithing

Pithing, also known as ikejime (or ikijime), involves sticking a sharp spike through the brain of the fish. If done properly, it can kill quickly, however, if the operator misses the brain, the results may be stressful for the fish. As with percussive stunning, spiking is used to kill one fish at a time and so is mainly used for large species such as tuna and salmon.

Shooting

Shooting large fish is also possible.

Electrical Stunning

Electricity can be more humane than alternatives if applied correctly. In addition to potentially producing unconsciousness quickly, stunning reduces the stress of restraint and being removed from water.

If electrical parameters are not optimized, electrical stunning may produce immobility without loss of consciousness, which is inhumane. There is little public data comparing optimal stun settings found by researchers with the settings used in commercial slaughter operations, so it is unknown how effective real-world stunning is. In addition, proper stun parameters vary significantly by species.

Electricity may introduce bleedspots, so proper settings are required.

Modern Systems

Systems have been developed to slaughter large numbers of fish whilst maintaining welfare standards.

A paper published by Jeff Lines and his collaborators in 2003 announced that stunning trout for 60 seconds in an electric field of 250 V/m r.m.s. with a sinusoidal waveform of 1,000 Hz rendered them permanently unconscious without degrading meat quality. A stunning system, called HS1, has been developed in accordance with Lines' study. The system first stuns fish and then keeps them unconscious, through electronarcosis, until death. The machine has been widely adopted in the UK, precessing an estimated 80% of all UK trout killed for meat. According to the Humane Slaughter Association's James Kirkwood: "Before ten years ago there was no way to humanely kill farmed fish en masse – they died slowly through suffocation when harvested from the water. This welfare benefit affects millions of fish."

Regulations

No welfare standards exist for the trillion or more fish harvested from the wild each year.

Since 2008, Norway has banned CO_2 stunning. By January 2010, 80% of Norwegian fish-slaughter facilities had switched to either percussive or electrical stunning.

Germany has banned use of salt or ammonia baths.

Fish Hatchery

Tanks in a shrimp hatchery.

A fish hatchery is a "place for artificial breeding, hatching and rearing through the early life stages of animals, finfish and shellfish in particular". Hatcheries produce larval and juvenile fish (and shellfish and crustaceans) primarily to support the aquaculture industry where they are transferred to on-growing systems i.e. fish farms to reach harvest size. Some species that are commonly raised in hatcheries include Pacific oysters, shrimp, Indian prawns, salmon, tilapia and scallops. The value of global aquaculture production is estimated to be US$98.4 billion in 2008 with China significantly dominating the market, however the value of aquaculture hatchery and nursery production has yet to be estimated. Additional hatchery production for small-scale domestic uses, which is particularly prevalent in South-East Asia or for conservation programmes, has also yet to be quantified.

There is much interest in supplementing exploited stocks of fish by releasing juveniles that may be wild caught and reared in nurseries before transplanting, or produced solely within a hatchery. Culture of finfish larvae has been utilised extensively in the United States in stock enhancement efforts to replenish natural populations. The U.S. Fish and Wildlife Service have established a National Fish Hatchery System to support the conservation of native fish species.

Purpose

Assynt Salmon hatchery, near Inchnadamph in the Scottish Highlands.

Hatcheries produce larval and juvenile fish and shellfish for transferral to aquaculture facilities where they are 'on-grown' to reach harvest size. Hatchery production confers three main benefits to the industry;

1. Out of season production

Consistent supply of fish from aquaculture facilities is an important market requirement. Broodstock conditioning can extend the natural spawning season and thus the supply of juveniles to farms. Supply can be further guaranteed by sourcing from hatcheries in the opposite hemisphere i.e. with opposite seasons.

2. Genetic improvement

Genetic modification is conducted in some hatcheries to improve the quality and yield of farmed species. Artificial fertilisation facilitates selective breeding programs which aim to improve production characteristics such as growth rate, disease resistance, survival, colour, increased fecundity and/or lower age of maturation. Genetic improvement can be mediated by selective breeding, via hybridization, or other genetic manipulation techniques.

3. Reduce dependence on wild-caught juveniles

In 2008 aquaculture accounted for 46% of total food fish supply, around 115 million tonnes. Although wild caught juveniles are still utilised in the industry, concerns over sustainability of extracting juveniles, and the variable timing and magnitude of natural spawning events, make hatchery production an attractive alternative to support the growing demands of aquaculture.

Production Steps

Manually stripping eggs

Juvenile salmon towards the end of their stay in a hatchery

Broodstock

Broodstock conditioning is the process of bringing adults into spawning condition by promoting the development of gonads. Broodstock conditioning can also extend spawning beyond natural spawning periods, or for production of species reared outside their natural geographic range with different environmental conditions. Some hatcheries collect wild adults and then bring them in for conditioning whilst others maintain a permanent breeding stock. Conditioning is achieved by holding broodstock in flow-through tanks at optimal conditions for light, temperature, salinity, flow rate and food availability (optimal levels are species specific). Another important aspect of broodstock conditioning is ensuring the production of high quality eggs to improve growth and survival of larvae by optimising the health and welfare of broodstock individuals. Egg quality is often determined by the nutritional condition of the mother. High levels of lipid reserves in particular are required to improve larval survival rates.

Spawning

Natural spawning can occur in hatcheries during the regular spawning season however where more control over spawning time is required spawning of mature animals can be induced by a variety of methods. Some of the more common methods are: Manual stripping: For shellfish, gonads are generally removed and gametes are extracted or washed free. Fish can be manually stripped of eggs and sperm by stroking the anaesthetised fish under the pectoral fins towards the anus causing gametes to freely flow out. Environmental manipulation: Thermal shock, where cool water is alternated with warmer water in flow-through tanks can induce spawning. Alternatively, if environmental cues that stimulate natural spawning are known, these can be mimicked in the tank e.g. changing salinity to simulate migratory behaviour. Many individuals can be induced to spawn this way, however this increases the likelihood of uncontrolled fertilisation occurring. Chemical injection: A number of chemicals can be used to induce spawning with various hormones being the most commonly used.

Fertilisation

Prior to fertilisation, eggs can be gently washed to remove wastes and bacteria that may contaminate cultures. Promoting cross-fertilisation between a large number of individuals is necessary to retain genetic diversity in hatchery produced stock. Batches of eggs are kept separate, fertilised with sperm obtained from several males and allowed to stand for an hour or two before samples are analyzed under a microscope to ensure high rates of fertilisation and to estimate numbers to be transferred to larval rearing tanks.

Larvae

Rearing larvae through the early life stages is conducted in nurseries which are generally closely associated with hatcheries for fish culture whilst it is common for shellfish nurseries to exist separately. Nursery culture of larvae to rear juveniles of a size suitable for transferral to on-growing facilities can be performed in a variety of different systems which may be entirely land-based, or larvae may be later transferred to sea-based rearing systems which reduce the need to supply feed. Juvenile survival is dependent on very high quality water conditions. Feeding is an important component of the rearing process. Although many species are able to grow on maternal reserves alone (lecithotrophy), most commercially produced species require feeding to optimise survival, growth, yield and juvenile quality. Nutritional requirements are species specific and also vary with larval stage. Carnivorous fish are commonly fed with live prey; rotifers are usually offered to early larvae due to their small size, progressing to larger *Artemia* nauplii or zooplankton. The production of live feed on-site or buying-in is one of the biggest costs for hatchery facilities as it is a labour-intensive process. The development of artificial feeds is targeted to reduce the costs involved in live feed production and increase the consistency of nutrition, however decreased growth and survival has been found with these alternatives.

Settlement of Shellfish

The hatchery production of shellfish also involves a crucial settling phase where free-swimming larvae settle out of the water onto a substrate and undergo metamorphosis if suitable conditions are found. Once metamorphosis has taken place the juveniles are generally known as spat, it is this phase which is then transported to on-growing facilities. Settlement behaviour is governed by a range of cues including substrate type, water flow, temperature, and the presence of chemical cues indicating the presence of adults, or a food source etc. Hatchery facilities therefore need to understand these cues to induce settlement and also be able to substitute artificial substrates to allow for easy handling and transportation with minimal mortality.

Hatchery Design

Multi-Species Fish and Invertebrate Breeding and Hatchery, (Oceanographic Marine Laboratory in Lucap, Alaminos, Pangasinan, Philippines, RMaTDeC,2011).

Hatchery designs are highly flexible and are tailored to the requirements of site, species produced, geographic location, funding and personal preferences. Many hatchery facilities are small and coupled to larger on-growing operations, whilst others may produce juveniles solely for sale. Very small-scale hatcheries are often utilized in subsistence farming to supply families or communities particularly in south-east Asia. A small-scale hatchery unit consists of larval rearing tanks, filters, live food production tanks and a flow through water supply. A generalized commercial scale hatchery would contain a broodstock holding and spawning area, feed culture facility, larval culture area, juvenile culture area, pump facilities, laboratory, quarantine area, and offices and bathrooms.

Expense

Labour is generally the largest cost in hatchery production making up more that 50% of total costs. Hatcheries are a business and thus economic viability and scale of production are vital considerations. The cost of production for stock-enhancement programmes is further complicated by the difficulty of assessing the benefits to wild populations from restocking activities.

Issues

Genetic

Hatchery facilities present three main problems in the field of genetics. The first is that maintenance of a small number of broodstock can cause inbreeding and potentially lead to inbreeding depression thus affecting the success of the facility. Secondly, hatchery reared juveniles, even from a fairly large broodstock, can have greatly reduced genetic diversity compared to wild populations (the situation is comparable to the founder effect). Such fish that escape from farms or are released for restocking purposes may adversely affect wild population genetics and viability. This is of particular concern where escaped fish have been actively bred or are otherwise genetically modified. The third key issue is that genetic modification of food items is highly undesirable for many people.

Fish Farms

Other arguments that surround fish farms such as the supplementation of feed from wild caught species, the prevalence of disease, fish welfare issues and potential effects on the environment are also issues for hatchery facilities.

Shark Finning

NOAA agent counting confiscated shark fins

Shark finning refers to the removal and delignening of shark fins while the remainder of the shark is discarded in the ocean. Sharks returned to the ocean without their fins are often still alive; unable to move effectively, they sink to the bottom of the ocean and die of suffocation or are eaten by other predators. Shark finning at sea enables fishing vessels to increase profitability and increase the number of sharks harvested, as they only have to store and transport the fins, by far the most profitable part of the shark. Some countries have banned this practice and require the whole shark to be brought back to port before removing the fins.

Shark finning increased since 1997 largely due to the increasing demand for shark fins for shark fin soup and traditional cures, particularly in China and its territories, and as a result of improved fishing technology and market economics. The International Union for Conservation of Nature's Shark Specialist Group say that shark finning is widespread, and that "the rapidly expanding and largely unregulated shark fin trade represents one of the most serious threats to shark populations worldwide". Estimates of the global value of the shark fin trade range from US$540 million to US$1.2 billion (2007). Shark fins are among the most expensive seafood products, commonly retailing at US$400 per kg. In the United States, where finning is prohibited, some buyers regard the whale shark and the basking shark as trophy species, and pay $10,000 to $20,000 for a fin.

The regulated global catch of sharks reported to the Food and Agriculture Organization of the United Nations has been stable in recent years at an annual average just over 500,000 tonnes. Additional unregulated and unreported catches are thought to be common.

Process

Nearly every fin of a shark is targeted for harvest, as highlighted in the diagram. The primary and secondary dorsal fins are removed from the top of the shark, plus its pectoral fins, and, in a single cutting motion, the pelvic fin, anal fin, and bottom portion of its caudal fin, or tail. The term "shark finning" specifically refers to the practice of removing the fins and discarding the carcass *while still at sea*. The removal of fins *on land* during catch processing is *not* considered shark finning.

Because the rest of the shark has little value relative to that of its fins, and because it is much bulkier, the finless and often still-living shark is thrown back into the sea to free space for more fins aboard the vessel. Shark species that are commonly finned are:

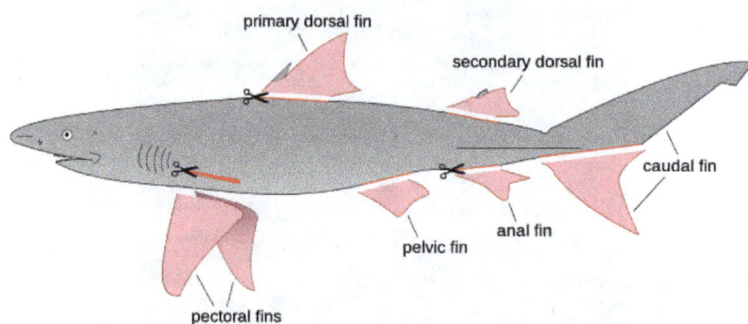

Highlights showing the typical targets of finning

- Blacktip (*Carcharhinus limbatus*)

- Blue (*Prionace glauca*) (a species of requiem shark)

- Bull (*Carcharhinus leucas*)

- Hammerhead (family Sphyrnidae)

- Porbeagle (*Lamna nasus*) (a species of mackerel shark)

- Mako (*Isurus oxyrinchus*)

- Sandbar (*Carcharhinus plumbeus*) (a species of requiem shark)

- Thresher (family Alopiidae)

- Tiger (*Galeocerdo cuvier*) (a species of requiem shark)

- Great white shark (*Carcharodon carcharias*)

Impacts

On Individual Sharks

The sand tiger shark is a large coastal shark that inhabits coastal waters worldwide. Its numbers have declined significantly, and it is now listed as a vulnerable species on the IUCN Red List.

Removal of a shark's fins prevents it from swimming. It is therefore incapable of hunting for prey or avoiding predators. Further, some species, known as *obligate ram ventilators*, lack the ability to pump water through their gills and must swim without rest; these species presumably asphyxiate if unable to move.

On Shark Populations

Some studies suggest 26 to 73 million sharks are harvested annually for fins. The annual median for the period from 1996 to 2000 was 38 million, which is nearly four times the number recorded by the Food and Agriculture Organization (FAO) of the United Nations, but considerably lower than the estimates of many conservationists. It has been reported that the global shark catch in 2012 was 100 million.

Sharks have a K-selection life history, which means that they tend to grow slowly, reach maturity at a larger size and a later age, and have low reproductive rates. These traits make them especially vulnerable to overfishing methods, such as shark finning. Recent studies suggest changes in abundance of apex predators may have cascading impacts on a variety of ecological processes.

Numbers of some shark species have dropped as much as 80% over the last 50 years. Some organizations claim that shark fishing or bycatch (the unintentional capture of species by other fisheries) is the reason for the decline in some species' populations and that the market for fins has very little impact – bycatch accounts for an estimated 50% of all sharks taken – others that the market for shark fin soup is the main reason for the decline.

On Other Populations

Sharks are apex predators and have extensive implications for marine systems an processes, particularly coral reefs. A report by WildAid on global threats to sharks further explains the importance of these animals.

Fins from the critically endangered sawfish (Pristidae) "are highly favored in Asian markets and are some of the most valuable shark fins". Sawfishes are now protected under the highest protection level of the Convention on International Trade in Endangered Species (CITES), Appendix I.

Vulnerability Of Sharks

On the IUCN Red List there are 39 species of elasmobranches (sharks and rays) listed as threatened species (Critically Endangered, Endangered or Vulnerable).

In 2013, the Convention on International Trade in Endangered Species of Wild Fauna and Flora (CITES) listed the vulnerability of sharks.

Appendix I, which lists animals that are threatened with extinction, lists

- Requiem sharks
- Hammerhead sharks
- Basking shark
- Mackerel sharks
- Whale shark

Appendix II, which lists animals that are not necessarily now threatened with extinction but that may become so unless trade is closely controlled, lists

- Basking shark (*Cetorhinus maximus*)
- Great white shark (*Carcharodon carcharias*)
- Whale shark (*Rhincodon typus*)

A further 5 species are listed, to come into effect in 2014 –

- Scalloped hammerhead (*Sphyrna lewini*)
- Great hammerhead shark (*Sphyrna mokarran*)
- Smooth hammerhead (*Sphyrna zygaena*)
- Porbeagle (*Lamna nasus*)
- Oceanic whitetip shark (*Carcharhinus longimanus*)

Opposition

A mural protesting shark finning in Wellington, New Zealand

The crew of the Sea Shepherd Conservation Society conservation vessel RV Ocean Warrior witnessed and photographed industrial-scale finning within Costa Rica's Cocos Island National Park protected marine area. The practice is featured in the documentary *Sharks: Stewards of the Reef,* which contains footage from Western Australia and Central America and also examines shark finning's cultural, financial and ecological impacts. Underwater photographer Richard Merritt witnessed finning of living sharks in Indonesia where he saw immobile finless sharks lying on the sea bed still alive below the fishing boat. Finning has been witnessed and filmed within a protected marine area in the Raja Ampat islands of Indonesia.

Animal welfare groups vigorously oppose finning on perceived moral grounds and also because it is one cause for the rapid decline of global shark populations.

Because of the lucrative profits and alleged size of the market, there are allegations of links to organized crime. Opponents also raise questions on the medical harm from the consumption of high levels of toxic mercury reportedly found in shark fins.

Shark fin fishing boat off the Galapagos, Ecuador

A third of fins imported to Hong Kong come from Europe. Spain is by far the largest supplier, providing between 2,000 and 5,000 metric tons a year. Norway supplies 39 metric tonnes, but Britain, France, Portugal and Italy are also major suppliers. Hong Kong handles at least 50% and possibly up to 80% of the world trade in shark fin, with the major suppliers being Europe, Taiwan, Indonesia, Singapore, United Arab Emirates, United States, Yemen, India, Japan, and Mexico. According to Giam's article, "Sharks are caught in virtually all parts of the world.... Despite the strongly declared objectives of the Fisheries Commission in Brussels, there are very few restrictions on fishing for sharks in European waters. The meat of dogfishes, smoothhounds, cat sharks, skates and rays is in high demand by European consumers.... The situation in Canada and the United States is similar: the blue shark is sought after as a sport fish while the porbeagle, mako and spiny dogfish are part of the commercial fishery.... The truth is this: Sharks will continue to be caught and killed on a wide scale by the more organized and sophisticated fishing nations. Targeting shark's fin soup will not stop this accidental catch. The fins from these catches will be thrown away or turned into animal feed and fertilizers if shark's fin soup is shunned."

In 2007, Canadian filmmaker and photographer Rob Stewart created a film, *Sharkwater*, which exposes the shark fin industry in detail. In March 2011, the VOA Special English service of the Voice of America broadcast a 15-minute science program on shark finning.

Reporting

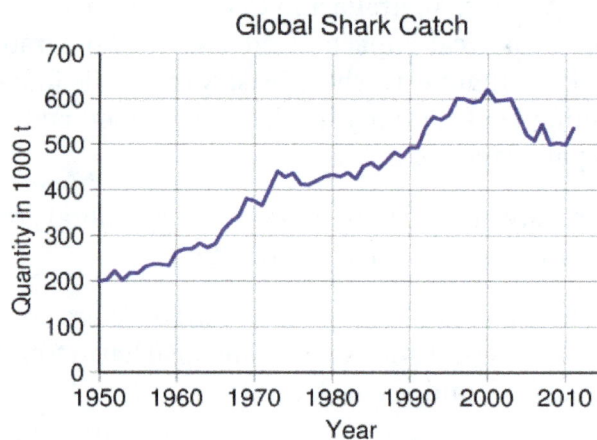

Global shark catch, since 1950. Source

According to Giam Choo Hoo – the longest serving member of The Convention on International Trade in Endangered Species of Wild Fauna and Flora Animals Committee, and a representative of the shark fin industry in Singapore – "The perception that it is common practice to kill sharks for only their fins – and to cut them off whilst the sharks are still alive – is wrong.... The vast majority of fins in the market are taken from sharks after their death."

However, researchers dispute this claim by pointing to the data: using a statistical analysis of shark fin industry trade data, a 2006 study estimated that between 26 and 73 million sharks are harvested each year worldwide. That figure, when converted to shark biomass, was three to four times higher than the catch recorded in Food and Agriculture Organization capture production statistics, the only global database of shark catches. According to the researchers, this discrepancy "may be attributable to factors... such as unrecorded shark landings, shark biomass recorded in [non-specific] categories, and/or a high frequency of shark finning and carcass disposal at sea." Simply put, they say that the industry is either under-reporting the sharks taken annually, or is frequently engaging in the practice of finning.

International Restrictions

Fresh shark fins drying on a pavement in Hong Kong

In 2013, 27 countries and the European Union had banned shark finning, however, international waters are unregulated. International fishing authorities are considering banning shark fishing (and finning) in the Atlantic Ocean and Mediterranean Sea. Finning is banned in the Eastern Pacific, but shark fishing and finning continues unabated in most of the Pacific and Indian Ocean. In countries such as Thailand and Singapore, public awareness advertisements on finning have reportedly reduced consumption by 25%.

There are four main categories of restrictions, as follows:

- Shark sanctuary (an area where shark fishing is entirely prohibited);

- Areas where sharks must be landed with fins attached;

- Areas where fin to body mass ratio-based regulations have been implemented;

- Areas where shark product trade regulations exist.

European Union

Shark finning was prohibited in the EU in 2003 (Regulation (EC) No 1185/2003).

In November 2011, the EC approved a rule that would require all EU-registered fishing boats to land only sharks which have retained all their fins. Because the legislation allowed fins to be removed on the boat and different body parts to be landed at different ports, the ban proved difficult to enforce. The EU Parliament's fisheries committee supported the EC's proposal to ban the separate landing of shark bodies and fins, however, the committee approved an amendment which allows fins to be removed on board a vessel.

On 19 March 2012, the Council of the EU adopted a general approach supporting the Commission's proposal to close the loopholes in the EU shark finning legislation by ensuring that all sharks were landed with their fins naturally attached without exception. It is believed that Spain and Portugal were the only EU Member States to raise objections to the Commission's proposal.

On 6 June 2013, the Council of the EU completed the final step to close loopholes in the EU shark finning ban. By adopting a "fins naturally attached" policy without exception, the EU has now effectively ended the practice of shark finning by EU vessels.

National Restrictions

Australia

Shark finning is not allowed in any tuna or billfish longline fishery, or in any Commonwealth fishery (i.e. federal waters ranges from 3–200 miles (4·8 to 321·9 km) offshore) taking sharks. Fins must be landed attached, and additional regulations apply in some states or territories. In New South Wales, sharks taken or any relevant portion of a shark taken may not be on board any vessel at any time (including after landing) without fins naturally attached.

Imported Products

In Australia, the export and import of wildlife and wildlife products is regulated under Part 13A of the federal Environment Protection and Biodiversity Conservation Act 1999 (EPBC Act), which is administered by the Department of Sustainability, Environment, Water, Population and Communities. Regulation applies equally to individuals, commercial organisations and not-for-profit organisations. CITES Appendix II shark specimens cannot be legally imported into Australia for personal or commercial purposes unless:

- The specimen is accompanied by a valid Australian CITES import permit (Australian import permits can be granted only if an overseas CITES export permit has been granted); or

- The specimen is accompanied by a valid certificate issued by the overseas CITES management authority confirming that the specimen was obtained before the species was listed on CITES (pre-CITES certificate); or

- The specimen is accompanied by an overseas CITES export permit or equivalent, is part of personal accompanied baggage and is intended for personal use and not for trade or sale.

No permits are required for the import of specimens obtained from shark species other than those listed above, however, to avoid seizure all products must be clearly labelled or have documentation certifying the species of origin.

Canada

Shark finning has been illegal in Canada since 1994, but importing fins from other regions without such regulations is allowed.

In late 2011, the city of Brantford, Ontario became the first city in Canada to pass new bylaws to ban the possession, sale or consumption of shark fin products. In that medium-sized city in which no restaurants exist which serve shark fin, there was no opposition to the ban, which was largely symbolic. Nevertheless, a handful of cities soon followed, notably Toronto, Calgary, Mississauga and several others in Southern Ontario:

- Brantford, Ontario 11 to 0 vote

- Oakville, Ontario 7 to 0 vote

- Mississauga, Ontario 11 to 0 vote (later repealed by Council on 8 May 2013)

- Toronto 38 to 4 vote (later overturned by court on 30 November 2012)

- Newmarket, Ontario 8 to 1 vote

- Calgary 13 to 2 vote

Markham and Richmond Hill opted not to bring forth the motion, suggesting that this issue is a federal matter. Chinese restaurants and businesses selling shark's fin opposed the ban, and in late 2011, suggested that they will challenge the by-laws before the courts once fines are imposed. When Toronto imposed steep fines, they did just that.

In late 2012, the Ontario Superior Court overturned Toronto's shark fin ban, ruling that the law as written was outside the powers of the city to impose without a "legitimate local purpose", and was therefore of "no force and effect." The judge accepted that the practice of shark finning was inhumane, but he did not agree with Toronto's justification of local purpose —– namely, that the consumption of shark fins may have an "adverse impact" on the health and safety of its residents and on the environmental well-being of the city. Toronto has served legal notice that it plans to appeal the court ruling.

On 1 December 2012, Ontario Superior Court Judge James Spence ruled that Toronto's ban was not valid. Members of Toronto's Chinese business community had also challenged that ban. Judge Spence had said the city does not have the power to enforce the ban. Toronto's mayor Rob Ford in September 2012 believed the ban was not the cities responsibility and so did not support it at that time.

On 27 March 2013 a private members bill to ban shark fin imports into Canada failed in the House of Commons. Shark finning was already illegal in Canadian waters, but there was no law to stop importing into Canada.

On 8 May 2013 Calgary's City Council decided to wait until December 2013 and recommended leaning away from a total ban and look for ethical sources of shark products. Alderman John Mar said there will be more time to discuss and engage and to look for other options. The new wording in the bylaw was meant to ban the sale, distribution and trade of shark fins, but not ban the possession and consumption. Canada's city of Vancouver councillor Kerry Jang was at Calgary's council meeting and said that it was not a cultural thing, and that even China and the Chinese government decided to phase out all shark fins from state banquets. He also mentioned that the wordings of the bylaws in Calgary and Toronto, which face legal problems with municipal jurisdiction, are trying to ban possession and consumption, but that is hard to enforce and regulate.

On 27 May 2013, against the wishes of the Shark Fin Free Calgary organization, Calgary City Council overturned the ban. There were protests against the ban from Calgary's Chinese community, and Calgary's city task force recommended against the ban. According to the article in The Calgary Herald, Calgary's Mayor Naheed Nenshi never wanted a full ban, even though he had voted for the ban the previous year.

China

Shark fins on display in a pharmacy in Yokohama, Japan

Protestors in Hong Kong, May 2012

NBA All-Star Chinese basketball player Yao Ming pledged to stop eating shark fin soup at a news conference on 2 August 2006. American basketball player Tracy McGrady, a teammate of Yao's, reportedly stated that he was impressed by the soup when he tried it for the first time, but was criticized by the Hong Kong branch of the World Wide Fund for Nature for his remark. The Australian naturalist Steve Irwin was known to walk out of Chinese restaurants if he saw shark fin soup on the menu. American chef Ken Hom sees the West doing little to protect stocks of cod and caviar-producing sturgeon despite the outcry over shark-finning, but he also stresses the wastefulness of harvesting only the fins.

Hong Kong

Disneyland Hong Kong removed shark fin soup from its wedding banquet menu after international pressure from environmental groups, who threatened to boycott its parks worldwide despite the high demand for the delicacy. The University of Hong Kong has banned shark fin soup on campus. The Peninsula Hotel banned shark fin in 2012.

Taiwan

Taiwan banned shark finning in 2011.

Malaysia

Malaysia was one of the top 10 importers and exporters of shark fins in the world between 2000 and 2009. The country caught 231,212 tonnes of sharks from 2002 to 2011, making it the eighth highest in the world and accounting for 2.9% of the global sharks caught during the same period.

In 2007, Malaysia's Natural Resources and Environment Ministry, Azmi Khalid, banned shark's fin soup from official functions committing to the Malaysian Nature Society (for conservation of shark species). In 2012, the Sabah Tourism, Culture and Environment Minister proposed an amendment to the Fisheries Act that would give force to setting up a shark sanctuary zone in Semporna and other shark populated areas in Sabah. This ban was put on hold pending the Federal Government's decision on the issue. In 2015, Agriculture and Agro-based Industry Minister, Ahmad Shabery Cheek, said that the ban of shark finning is "unnecessary" as the finning industry does not exist in Malaysia He went on further to say "Sharks are normally caught by accident when they enter the fish nets along with the other fishes."

New Zealand

The great white sharks have been given full protection in the territorial waters of New Zealand but shark finning is legal on other shark species if the shark is dead. The Royal Forest and Bird Protection Society of New Zealand are campaigning to raise awareness of shark finning and a number of foodies have fronted the campaign.

At 10 November 2013, an announcement was made by the conservation minister proposing that shark finning will be outlawed in New Zealand waters.

Palau

In 2009, the Republic of Palau created the world's first shark sanctuary. It is illegal to catch sharks within Palau's EEZ, which covers an area of 230,000 square miles (600,000 km²). This is an area about the size of France. President Johnson Toribiong also called for a ban on global shark finning, stating: "These creatures are being slaughtered and are perhaps at the brink of extinction unless we take positive action to protect them."

Singapore

Leading Singapore-based supermarket chain Cold Storage, has joined the World Wide Fund for

Nature Singapore Sustainable seafood Group and agreed to stop selling all shark fin and shark products in its 42 outlets across the country. The supermarket is a subsidiary of Dairy Farm, a leading pan-Asian food retailer that operates more than 5,300 outlets and employs some 80,000 people in the Asia-Pacific region. It is the first supermarket in Singapore to implement a no shark fins policy.

The largest supermarket chain in Singapore, NTUC Fairprice and hypermarket Carrefour will also be banning all shark fin products from its outlets before April 2012.

United States

Bill Clinton signed the Shark Finning Prohibition Act of 2000 (SFPA), which banned finning on any fishing vessel within United States territorial waters, and on all U.S.-flagged fishing vessels in international waters. Additionally, shark fins could not be imported into the United States without the associated carcass. In 1991, the percentage of sharks killed by U.S. longline fisheries in the Pacific Ocean for finning was approximately 3%. By 1998, that percentage had grown to 60%. Between 1991 and 1998, the number of sharks retained by the Hawaii-based swordfish and tuna longline fishery had increased from 2,289 to 60,857 annually, and by 1998, an estimated 98% of these sharks were killed for their fins.

The *King Diamond II* was seized in 2002 while carrying over 32 tons (29 metric tonnes) of shark fins that were harvested from about 30,000 sharks

In 2002, in an apparent early success in stopping the shark fin trade, the United States intercepted and seized the *King Diamond II*, a U.S.-flagged, Hong Kong-based vessel bound for Guatemala. The vessel was carrying 32.3 tons (29.3 tonnes) of baled shark fins – representing the fins of an estimated 30,000 sharks – making it the largest quantity of shark fins ever seized. This seizure was reversed in court six years later: in *United States v. Approximately 64,695 Pounds of Shark Fins*, the Ninth Circuit Court of Appeals held that the SFPA did not cover the seized fins in this case. Judge Stephen Reinhardt found that the *King Diamond II* did not meet the statute's definition of a fishing vessel, since it had merely bought the fins at sea and had not aided or assisted the vessels that had caught the sharks.

As a result, in January 2011, President Barack Obama signed the Shark Conservation Act into law to close the loopholes. Specifically, the new law prohibits any boat to carry shark fins without the corresponding number and weight of carcasses, and all sharks must be brought to port with their fins attached.

In 2010, Hawaii became the first state to ban the possession, sale and distribution of shark fins. The law became effective on 1 July 2011. Similar laws have been enacted in the states of Washington, Oregon, California, the territory of Guam and the Commonwealth of the Northern Mariana Islands. California governor Jerry Brown cited the cruelty of finning and potential threats to the environment and commercial fishing in signing the bill. Opponents charged the ban was discriminatory against Chinese, the main consumers of shark fin soup, when federal laws already banned the practice of finning. Whole sharks would still be legally fished, but the fins could no longer be sold.

In 2012, legislators in the New York State Assembly, including Grace Meng, introduced a similar bill. While New York was not the only Eastern state considering a ban, passage there would be significant since its Chinese American communities in Chinatown, Manhattan and Flushing make New York the major importer of shark fins in the East. Meng admitted that while she loved shark fin soup, "it's important to be responsible citizens." Younger Chinese Americans in New York did not consider it an important part of their culture. "It's only the elderly who want it: when their grandkids get married, they want the most expensive stuff, like an emperor," said one waiter at a Chinese restaurant. Many businesses that sold fins had stopped placing new orders, expecting a ban would be passed.

Bycatch

Shrimp bycatch

Bycatch, in the fishing industry, is a fish or other marine species that is caught unintentionally while catching certain target species and target sizes of fish, crabs etc. Bycatch is either of a different species, the wrong sex, or is undersized or juvenile individuals of the target species. The term "bycatch" is also sometimes used for untargeted catch in other forms of animal harvesting or collecting.

In 1997, the Organisation for Economic Co-operation and Development (OECD) defined bycatch as "total fishing mortality, excluding that accounted directly by the retained catch of target species". Bycatch contributes to fishery decline and is a mechanism of overfishing for unintentional catch.

The fisherman bycatch issue originated due to the "mortality of dolphins in tuna nets in the 1960s"

There are at least four different ways the word "bycatch" is used in fisheries:

- Catch which is retained and sold but which is not the target species for the fishery
- Species/sizes/sexes of fish which fishermen discard
- Non-target fish, whether retained and sold or discarded
- Unwanted invertebrate species, such as echinoderms and non-commercial crustaceans, and various vulnerable species groups, including seabirds, sea turtles, marine mammals and elasmobranchs (sharks and their relatives).

Examples

Recreational Fishing

Given the popularity of recreational fishing throughout the world, a small local study in the US in 2013 suggested that discards may be an important unmonitored source of fish mortality.

Shrimp Trawling

Double-rigged shrimp trawler hauling in the nets

Shrimp bycatch

The highest rates of incidental catch of non-target species are associated with tropical shrimp trawling. In 1997, the Food and Agriculture Organization of the United Nations (FAO) documented the estimated bycatch and discard levels from shrimp fisheries around the world. They found discard rates (bycatch to catch ratios) as high as 20:1 with a world average of 5.7:1.

Shrimp trawl fisheries catch 2% of the world total catch of all fish by weight, but produce more than one-third of the world total bycatch. American shrimp trawlers produce bycatch ratios between 3:1 (3 bycatch:1 shrimp) and 15:1(15 bycatch:1 shrimp).

Trawl nets in general, and shrimp trawls in particular, have been identified as sources of mortality for cetacean and finfish species. When bycatch is discarded (returned to the sea), it is often dead or dying.

Tropical shrimp trawlers often make trips of several months without coming to port. A typical haul may last 4 hours after which the net is pulled in. Just before it is pulled on board the net is washed by zigzagging at full speed. The contents are then dumped on deck and are sorted. An average of 5.7:1 means that for every kilogram of shrimp there are 5.7 kg of bycatch. In tropical inshore waters the bycatch usually consists of small fish. The shrimps are frozen and stored on-board; the bycatch is discarded.

Recent sampling in the South Atlantic rock shrimp fishery found 166 species of finfish, 37 crustacean species, and 29 other species of invertebrate among the bycatch in the trawls. Another sampling of the same fishery over a two-year period found that rock shrimp amounted to only 10% of total catch weight. Iridescent swimming crab, dusky flounder, inshore lizardfish, spot, brown shrimp, longspine swimming crabs, and other bycatch made up the rest.

Despite the use of bycatch reduction devices, the shrimp fishery in the Gulf of Mexico removes about 25–45 million red snapper annually as bycatch, nearly one half the amount taken in directed recreational and commercial snapper fisheries.

Cetacean

Group of Fraser's dolphins.

Cetaceans, such as dolphins, porpoises, and whales, can be seriously affected by entanglement in fishing nets and lines, or direct capture by hooks or in trawl nets. Cetacean bycatch is increasing in intensity and frequency. In some fisheries, cetaceans are captured as bycatch but then retained because of their value as food or bait. In this fashion, cetaceans can become a target of fisheries.

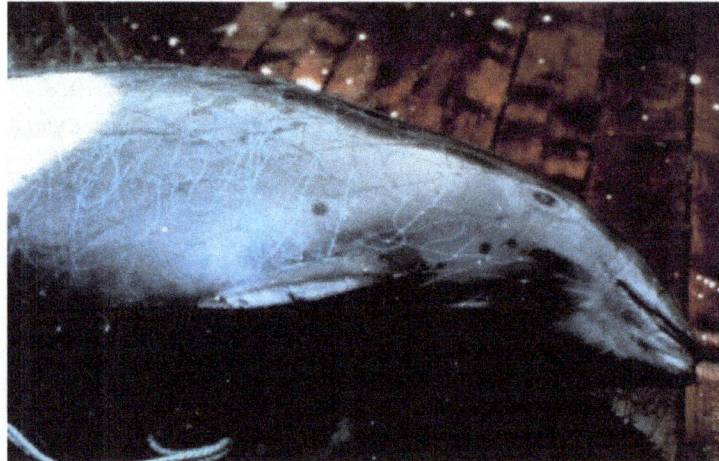

A Dall's porpoise caught in a fishing net

One example of bycatch is dolphins caught in tuna nets. As dolphins are mammals and do not have gills they may drown while stuck in nets underwater. This bycatch issue has been one of the reasons of the growing ecolabelling industry, where fish producers mark their packagings with disclaimers such as "dolphin friendly" to reassure buyers. However, "dolphin friendly" does not mean that dolphins were not killed in the production of a particular tin of tuna, but that the fleet which caught the tuna did not *specifically* target a feeding pod of dolphins, but relied on other methods to spot tuna schools.

Albatross

This black-browed albatross has been hooked on a long-line.

Of the 21 albatross species recognised by IUCN on their Red List, 19 are threatened, and the other two are *near threatened*. Two species are considered critically endangered: the Amsterdam albatross and the Chatham albatross. One of the main threats is commercial long-line fishing, because the albatrosses and other seabirds which readily feed on offal are attracted to the set bait, become hooked on the lines and drown. An estimated 100,000 albatross per year are killed in this fashion. Unregulated pirate fisheries exacerbate the problem.

Sea Turtles

Loggerhead sea turtle

Sea turtles, already critically endangered, have been killed in large numbers in shrimp trawl nets. Estimates indicate that thousands of Kemp's ridley, loggerhead, green and leatherback sea turtles are caught in shrimp trawl fisheries in the Gulf of Mexico and the US Atlantic annually The speed and length of the trawl method is significant because, "for a tow duration of less than 10 minutes, the mortality rate for sea turtles is less than one percent, whereas for tows greater than sixty minutes the mortality rate rapidly increases to fifty to one hundred percent"

Sea turtles can sometimes escape from the trawls. In the Gulf of Mexico, the Kemp's ridley turtles recorded most interactions, followed in order by loggerhead, green, and leatherback sea turtles. In the US Atlantic, the interactions were greatest for loggerheads, followed in order by Kemp's ridley, leatherback, and green sea turtles.

Mitigation

A turtle excluder device

Concern about bycatch has led fishermen and scientists to seek ways of reducing unwanted catch. There are two main approaches.

One approach is to ban fishing in areas where bycatch is unacceptably high. Such area closures can be permanent, seasonal, or for a specific period when a bycatch problem is registered. Temporary area closures are common in some bottom-trawl fisheries where undersized fish or non-target species are caught unpredictably. In some cases fishermen are required to relocate when a bycatch problem occurs.

The other approach is alternative fishing gear. A technically simple solution is to use nets with a larger mesh size, allowing smaller species and smaller individuals to escape. However, this usually requires replacing the existing gear. In other cases, it is possible to modify gear. The "Bycatch Reduction Device" (BRD) and the Nordmore grate are net modifications that help fish escape from shrimp nets.

BRDs allow many commercial finfish species to escape. The US government has approved BRDs that reduce finfish bycatch by 30%. Spanish mackerel and weakfish bycatch in the South Atlantic was reduced by 40%. However, recent surveys suggest BRDs may be less effective than previously thought. A rock shrimp fishery off Florida found the devices did not exclude 166 species of fish, 37 crustacean species, and 29 species of other invertebrates.

In 1978, the National Marine Fisheries Service (NMFS) started to develop turtle excluder devices (TEDs). A TED uses a grid which deflects turtles and other big animals, so they exit from the trawl net through an opening above the grid. US shrimp trawlers and foreign fleets which market shrimp in the US are required to use TEDs. Not all nations enforce the use of TEDs.

For the most part, when they are used, TEDs have been successful reducing sea turtle bycatch. However, they are not completely effective, and some turtles are still captured. NMFS certifies TED designs if they are 97% effective. In heavily trawled areas, the same sea turtle may pass repeatedly through TEDs. Recent studies indicate recapture rates of twenty percent or more, but it is not clear how many turtles survive the escape process.

The size selectivity of trawl nets is controlled by the size of the net openings, especially in the "cod end". The larger the openings, the more easily small fish can escape. The development and testing of modifications to fishing gear to improve selectivity and decrease impact is called "conservation engineering."

Seabirds with longline fishing vessel

Longline fishing is controversial in some areas because of by-catch. Mitigation methods have been successfully implemented in some fisheries. These include:

- weights to sink the lines quickly

- streamer lines to scare birds away from baited hooks while deploying the lines

- setting lines only at night with minimal ship lighting (to avoid attracting birds)

- limiting fishing seasons to the southern winter (when most seabirds are not feeding young)

- not discharging offal while setting lines.

However, gear modifications do not eliminate by-catch of many species. In March 2006, the Hawai'i longline swordfish fishing season was closed due to excessive loggerhead sea turtle by-catch after being open only a few months, despite using modified circle hooks.

One solution that Norway came up with to reduce bycatch is to, "adopt a 'no discards' policy". This means that the fishermen must keep everything they catch. This policy has helped to, "encourage [bycatch] research", which, in turn has helped "encourage behavioral changes in fishers" and "reduce the waste of life" as well.

Alternative to Release

Some fisheries retain bycatch, rather than throwing the fish back into the ocean. Sometimes bycatch are sorted and sold as food, especially in Asia, Africa and Latin America where cost of labour is cheaper. Bycatch can also be sold in frozen bags as "assorted seafood" or "seafood medley" at cheaper prices. Bycatch can be converted into fish hydrolysate (ground up fish carcasses) for use as a soil amendment in organic agriculture or it can be used as an ingredient in fish meal. In Southeast Asia bycatch is sometimes used as a raw material for fish sauce production. Bycatch is also commonly de-boned, de-shelled, ground and blended into fish paste or moulded into fish cakes (surimi) and sold either fresh (for domestic use) or frozen (for export). This is commonly the case in Asia or by Asian fisheries. Sometimes bycatch are sold to fish farms to feed farmed fish, especially in Asia.

If bycatch is quickly released, predators and scavengers may consume it.

Non-fisheries Bycatch

The term "bycatch" is used also in contexts other than fisheries. Examples are insect collecting with pitfall traps or flight interception traps for either financial, controlling or scientific purposes (where the bycatch may either be small vertebrates or untargeted insects) and control of introduced vertebrates which have become pest species like the muskrat in Europe (where the bycatch in traps may be e.g. European mink or waterfowl).

Fish Diseases and Parasites

Like humans and other animals, fish suffer from diseases and parasites. Fish defences against disease are specific and non-specific. Non-specific defences include skin and scales, as well as the mucus layer secreted by the epidermis that traps microorganisms and inhibits their growth. If pathogens breach these defences, fish can develop inflammatory responses that increase the flow of blood to infected areas and deliver white blood cells that attempt to destroy the pathogens.

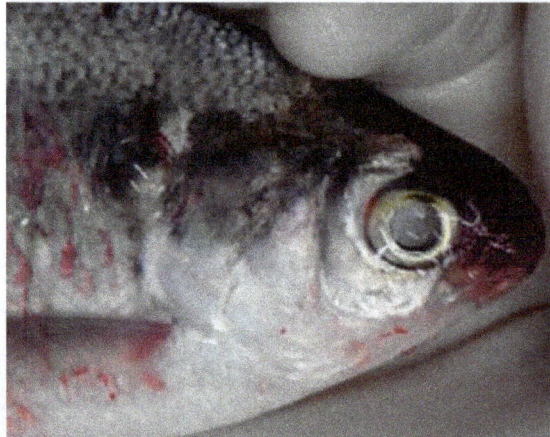

This gizzard shad has VHS, a deadly infectious disease which causes bleeding. It afflicts over 50 species of freshwater and marine fish in the northern hemisphere.

This flatfish *Limanda limanda* has an outgrowth called a xenoma. It is caused by a microsporidian fungal parasite in its intestines.

Specific defences are specialised responses to particular pathogens recognised by the fish's body, that is adaptative immune responses. In recent years, vaccines have become widely used in aquaculture and ornamental fish, for example vaccines for furunculosis in farmed salmon and koi herpes virus in koi.

Some commercially important fish diseases are VHS, ich and whirling disease.

Disease

All fish carry pathogens and parasites. Usually this is at some cost to the fish. If the cost is sufficiently high, then the impacts can be characterised as a disease. However disease in fish is not understood well. What is known about fish disease often relates to aquaria fish, and more recently, to farmed fish.

Disease is a prime agent affecting fish mortality, especially when fish are young. Fish can limit the impacts of pathogens and parasites with behavioural or biochemical means, and such fish have reproductive advantages. Interacting factors result in low grade infection becoming fatal diseases. In particular, things that causes stress, such as natural droughts or pollution or predators, can precipitate outbreak of disease.

A veterinarian gives an injection to a goldfish

Disease can also be particularly problematic when pathogens and parasites carried by introduced species affect native species. An introduced species may find invading easier if potential predators and competitors have been decimated by disease.

Pathogens which can cause fish diseases comprise:

- viral infections

- bacterial infections, such as *Pseudomonas fluorescens* leading to fin rot and fish dropsy

- fungal infections

- water mould infections, such as *Saprolegnia* sp.

- metazoan parasites, such as copepods

- unicellular parasites, such as *Ichthyophthirius multifiliis* leading to ich

- Certain parasites like Helminths for example *Eustrongylides*

Parasites

Parasites in fish are a common natural occurrence. Parasites can provide information about host population ecology. In fisheries biology, for example, parasite communities can be used to distinguish distinct populations of the same fish species co-inhabiting a region. Additionally, parasites possess a variety of specialized traits and life-history strategies that enable them to colonize hosts. Understanding these aspects of parasite ecology, of interest in their own right, can illuminate parasite-avoidance strategies employed by hosts.

The isopod Anilocra gigantea parasitising the snapper *Pristipomoides filamentosus*

Cymothoa exigua is a parasitic crustacean which enters a fish through its gills and destroys the fish's tongue.

Usually parasites (and pathogens) need to avoid killing their hosts, since extinct hosts can mean extinct parasites. Evolutionary constraints may operate so parasites avoid killing their hosts, or the natural variability in host defensive strategies may suffice to keep host populations viable. Parasite infections can impair the courtship dance of male threespine sticklebacks. When that happens, the females reject them, suggesting a strong mechanism for the selection of parasite resistance."

However, not all parasites want to keep their hosts alive, and there are parasites with multistage life cycles who go to some trouble to kill their host. For example, some tapeworms make some fish behave in such a way that a predatory bird can catch it. The predatory bird is the next host for the parasite in the next stage of its life cycle. Specifically, the tapeworm *Schistocephalus solidus* turns infected threespine stickleback white, and then makes them more buoyant so that they splash along at the surface of the water, becoming easy to see and easy to catch for a passing bird.

Parasites can be internal (endoparasites) or external (ectoparasites). Some internal fish parasites are spectacular, such as the philometrid nematode *Philometra fasciati* which is parasitic in the ovary of female Blacktip grouper; the adult female parasite is a red worm which can reach up to 40 centimetres in length, for a diameter of only 1.6 millimetre; the males are tiny. Other internal parasites are found living inside fish gills, include encysted adult didymozoid trematodes, a few tricho-somoidid nematodes of the genus *Huffmanela*, including *Huffmanela ossicola* which lives within the gill bone, and the encysted parasitic turbellarian *Paravortex*. Various protists and Myxosporea are also parasitic on gills, where they form cysts.

Fish gills are also the preferred habitat of many external parasites, attached to the gill but living out of it. The most common are monogeneans and certain groups of parasitic copepods, which can be extremely numerous. Other external parasites found on gills are leeches and, in seawater, larvae of gnathiid isopods. Isopod fish parasites are mostly external and feed on blood. The larvae of the Gnathiidae family and adult cymothoidids have piercing and sucking mouthparts and clawed limbs adapted for clinging onto their hosts. *Cymothoa exigua* is a parasite of various marine fish. It causes the tongue of the fish to atrophy and takes its place in what is believed to be the first instance discovered of a parasite functionally replacing a host structure in animals.

Other parasitic disorders, include *Gyrodactylus salaris*, *Ichthyophthirius multifiliis*, cryptocaryon, velvet disease, *Brooklynella hostilis*, Hole in the head, *Glugea*, *Ceratomyxa shasta*, *Kudoa thyrsites*, *Tetracapsuloides bryosalmonae*, *Cymothoa exigua*, leeches, nematode, flukes, carp lice and salmon lice.

Although parasites are generally considered to be harmful, the eradication of all parasites would not necessarily be beneficial. Parasites account for as much as or more than half of life's diversity; they perform an important ecological role (by weakening prey) that ecosystems would take some time to adapt to; and without parasites organisms may eventually tend to asexual reproduction, diminishing the diversity of sexually dimorphic traits. Parasites provide an opportunity for the transfer of genetic material between species. On rare, but significant, occasions this may facilitate evolutionary changes that would not otherwise occur, or that would otherwise take even longer.

Below are some life cycles of fish parasites:

Life cycle of the fish parasite *Ichthyophthirius multifiliis*, commonly called ich

Life cycle of the parasitic fluke *Clinostomum marginatum*, commonly called the yellow grub

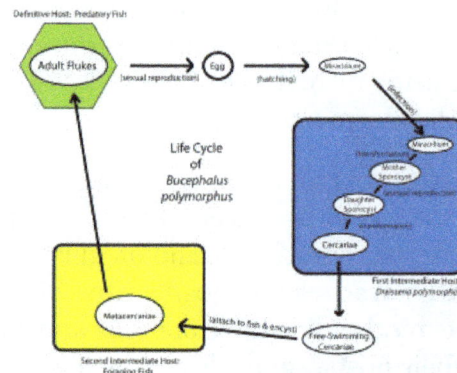

Life cycle of the digenean *Bucephalus polymorphus*

Cleaner Fish

Two cleaner wrasses, *Labroides phthirophagus*, servicing a goatfish, *Mulloidichthys flavolineatus*

Some fish take advantage of cleaner fish for the removal of external parasites. The best known of these are the Bluestreak cleaner wrasses of the genus *Labroides* found on coral reefs in the Indian Ocean and Pacific Ocean. These small fish maintain so-called "cleaning stations" where other fish, known as hosts, will congregate and perform specific movements to attract the attention of the cleaner fish. Cleaning behaviours have been observed in a number of other fish groups, including an interesting case between two cichlids of the same genus, *Etroplus maculatus*, the cleaner fish, and the much larger *Etroplus suratensis*, the host.

More than 40 species of parasites may reside on the skin and internally of the ocean sunfish, motivating the fish to seek relief in a number of ways. In temperate regions, drifting kelp fields harbour cleaner wrasses and other fish which remove parasites from the skin of visiting sunfish. In the tropics, the *mola* will solicit cleaner help from reef fishes. By basking on its side at the surface, the sunfish also allows seabirds to feed on parasites from their skin. Sunfish have been reported to breach more than ten feet above the surface, possibly as another effort to dislodge parasites on the body.

Mass Die Offs

Some diseases result in mass die offs. One of the more bizarre and recently discovered diseases produces huge fish kills in shallow marine waters. It is caused by the ambush predator dinoflagellate *Pfiesteria piscicida*. When large numbers of fish, like shoaling forage fish, are in confined situations such as shallow bays, the excretions from the fish encourage this dinoflagellate, which is not normally toxic, to produce free-swimming zoospores. If the fish remain in the area, continuing to provide nourishment, then the zoospores start secreting a neurotoxin. This toxin results in the fish developing bleeding lesions, and their skin flakes off in the water. The dinoflagellates then eat the blood and flakes of tissue while the affected fish die. Fish kills by this dinoflagellate are common, and they may also have been responsible for kills in the past which were thought to have had other causes. Kills like these can be viewed as natural mechanisms for regulating the population of exceptionally abundant fish. The rate at which the kills occur increases as organically polluted land runoff increases.

Wild Salmon

Henneguya salminicola, a parasite commonly found in the flesh of salmonids on the West Coast of Canada. Coho salmon

According to Canadian biologist Dorothy Kieser, protozoan parasite *Henneguya salminicola* is commonly found in the flesh of salmonids. It has been recorded in the field samples of salmon returning to the Queen Charlotte Islands. The fish responds by walling off the parasitic infection into a number of cysts that contain milky fluid. This fluid is an accumulation of a large number of parasites.

Henneguya and other parasites in the myxosporean group have a complex lifecycle where the salmon is one of two hosts. The fish releases the spores after spawning. In the *Henneguya* case, the spores enter a second host, most likely an invertebrate, in the spawning stream. When juvenile salmon out-migrate to the Pacific Ocean, the second host releases a stage infective to salmon. The parasite is then carried in the salmon until the next spawning cycle. The myxosporean parasite that causes whirling disease in trout, has a similar lifecycle. However, as opposed to whirling disease, the *Henneguya* infestation does not appear to cause disease in the host salmon — even heavily infected fish tend to return to spawn successfully.

According to Dr. Kieser, a lot of work on *Henneguya salminicola* was done by scientists at the Pacific Biological Station in Nanaimo in the mid-1980s, in particular, an overview report which states that "the fish that have the longest fresh water residence time as juveniles have the most noticeable infections. Hence in order of prevalence coho are most infected followed by sockeye, chinook, chum and pink." As well, the report says that, at the time the studies were conducted, stocks from the middle and upper reaches of large river systems in British Columbia such as Fraser, Skeena, Nass and from mainland coastal streams in the southern half of B.C. "are more likely to have a low prevalence of infection." The report also states "It should be stressed that *Henneguya*, economically deleterious though it is, is harmless from the view of public health. It is strictly a fish parasite that cannot live in or affect warm blooded animals, including man".

Sample of pink salmon infected with Henneguya salminicola, caught off the Queen Charlotte Islands, Western Canada in 2009

According to Klaus Schallie, Molluscan Shellfish Program Specialist with the Canadian Food Inspection Agency, "*Henneguya salminicola* is found in southern B.C. also and in all species of salmon. I have previously examined smoked chum salmon sides that were riddled with cysts and some sockeye runs in Barkley Sound (southern B.C., west coast of Vancouver Island) are noted for their high incidence of infestation."

Sea lice, particularly *Lepeophtheirus salmonis* and a variety of *Caligus* species, including *Caligus clemensi* and *Caligus rogercresseyi*, can cause deadly infestations of both farm-grown and wild salmon. Sea lice are ectoparasites which feed on mucous, blood, and skin, and migrate and latch onto the skin of wild salmon during free-swimming, planktonic *naupli* and *copepodid* larval stages, which can persist for several days. Large numbers of highly populated, open-net salmon farms can create exceptionally large concentrations of sea lice; when exposed in river estuaries containing large numbers of open-net farms, many young wild salmon are infected, and do not survive as a result. Adult salmon may survive otherwise critical numbers of sea lice, but small, thin-skinned juvenile salmon migrating to sea are highly vulnerable. On the Pacific coast of Canada, the louse-induced mortality of pink salmon in some regions is commonly over 80%.

Farmed Salmon

Atlantic salmon

In 1972, Gyrodactylus salaris, also called salmon fluke, a monogenean parasite, spread from Norwegian hatcheries to wild salmon, and devastated some wild salmon populations.

In 1984, infectious salmon anemia (ISAv) was discovered in Norway in an Atlantic salmon hatchery. Eighty percent of the fish in the outbreak died. ISAv, a viral disease, is now a major threat to the viability of Atlantic salmon farming. It is now the first of the diseases classified on List One of the European Commission's fish health regime. Amongst other measures, this requires the total eradication of the entire fish stock should an outbreak of the disease be confirmed on any farm. ISAv seriously affects salmon farms in Chile, Norway, Scotland and Canada, causing major economic losses to infected farms. As the name implies, it causes severe anemia of infected fish. Unlike mammals, the red blood cells of fish have DNA, and can become infected with viruses. The fish develop pale gills, and may swim close to the water surface, gulping for air. However, the disease can also develop without the fish showing any external signs of illness, the fish maintain a normal appetite, and then they suddenly die. The disease can progress slowly throughout an infected farm and, in the worst cases, death rates may approach 100 percent. It is also a threat to the dwindling stocks of wild salmon. Management strategies include developing a vaccine and improving genetic resistance to the disease.

In the wild, diseases and parasites are normally at low levels, and kept in check by natural predation on weakened individuals. In crowded net pens they can become epidemics. Diseases and parasites also transfer from farmed to wild salmon populations. A recent study in British Columbia links the spread of parasitic sea lice from river salmon farms to wild pink salmon in the same river." The European Commission (2002) concluded "The reduction of wild salmonid abundance is also linked to other factors but there is more and more scientific evidence establishing a direct link between the number of lice-infested wild fish and the presence of cages in the same estuary." It is reported that wild salmon on the west coast of Canada are being driven to extinction by sea lice from nearby salmon farms. Antibiotics and pesticides are often used to control the diseases and parasites.

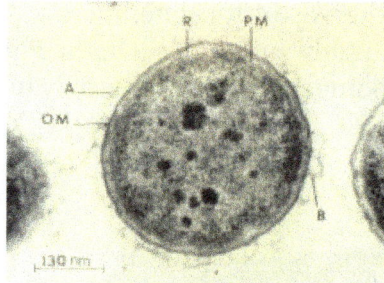

Aeromonas salmonicida, a Gram-negative bacteria, causes the disease furunculosis in marine and freshwater fish.

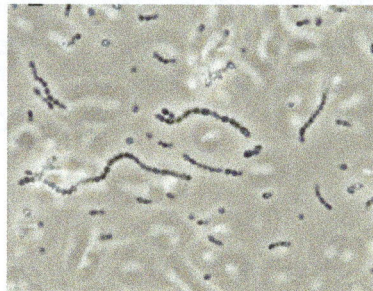

Streptococcus iniae, a Gram-positive, sphere-shaped bacteria caused losses in farmed marine and freshwater finfish of US$100 million in 1997.

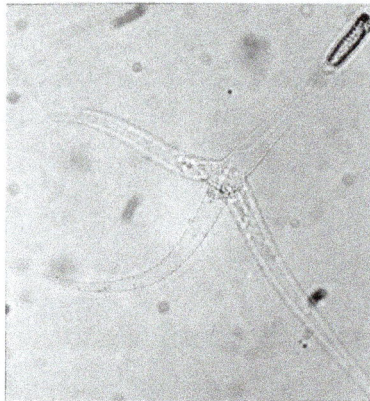

Myxobolus cerebralis, a myxosporean parasite, causes *whirling disease* in farmed salmon and trout and also in wild fish populations.

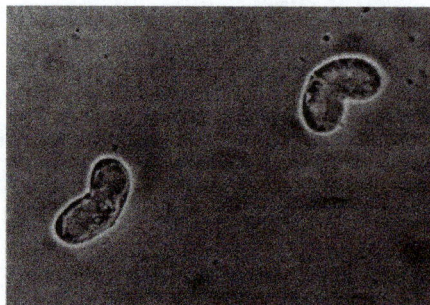

Ceratomyxa shasta, another myxosporean parasite, infects salmonid fish on the Pacific coast of North America.

Coral Reef Fish

Philometra fasciati (Nematoda), a roundworm parasitic of the ovary of the blacktip grouper

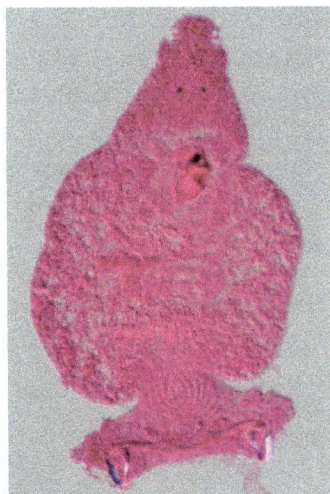

Monogenean parasite on the gill of a grouper

Coral reef fish are characterized by high biodiversity. As a consequence parasites of coral reef fish show tremendous variety. Parasites of coral reef fish include nematodes, Platyhelminthes (cestodes, digeneans, and monogeneans), leeches, parasitic crustaceans such as isopods and copepods, and various microorganisms such as myxosporidia and microsporidia. Some of these fish parasites have heteroxenous life cycles (i.e. they have several hosts) among which sharks (certain cestodes) or molluscs (digeneans). The high biodiversity of coral reefs increases the complexity of the interactions between parasites and their various and numerous hosts. Numerical estimates of parasite biodiversity have shown that certain coral fish species have up to 30 species of parasites. The mean number of parasites per fish species is about ten. This has a consequence in term of co-extinction. Results obtained for the coral reef fish of New Caledonia suggest that extinction of a coral reef fish species of average size would eventually result in the co-extinction of at least ten species of parasites.

Aquarium Fish

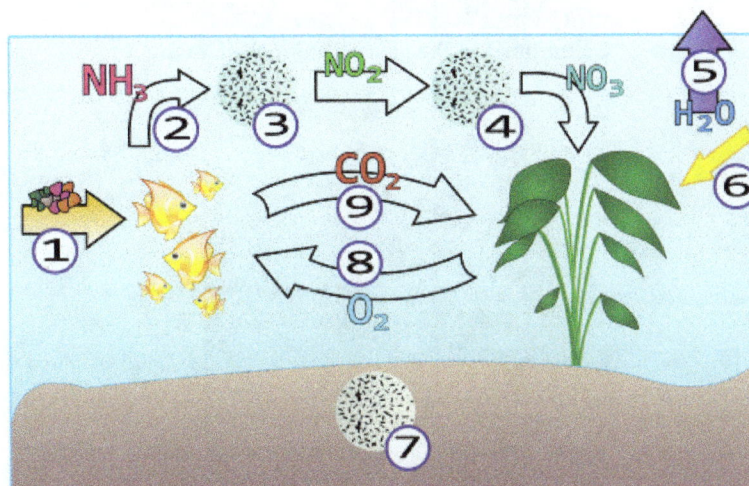

Nitrogen cycle in a common aquarium.

Ornamental fish kept in aquariums are susceptible to numerous diseases.

In most aquarium tanks, the fish are at high concentrations and the volume of water is limited. This means that communicable diseases can spread rapidly to most or all fish in a tank. An improper nitrogen cycle, inappropriate aquarium plants and potentially harmful freshwater invertebrates can directly harm or add to the stresses on ornamental fish in a tank. Despite this, many diseases in captive fish can be avoided or prevented through proper water conditions and a well-adjusted ecosystem within the tank. Ammonia poisoning is a common disease in new aquariums, especially when immediately stocked to full capacity.

Due to their generally small size and the low cost of replacing diseased or dead aquarium fish, the cost of testing and treating diseases is often seen as more trouble than the value of the fish.

Goldfish with dropsy

Columnaris in the gill of a chinook salmon

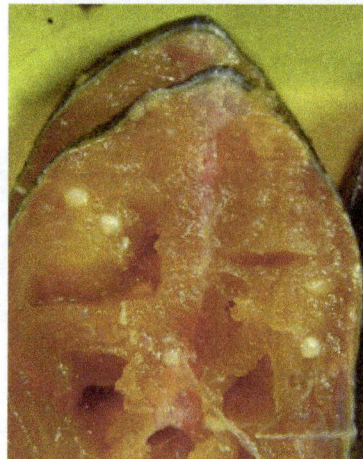

The parasite *Henneguya zschokkei* in salmon beard

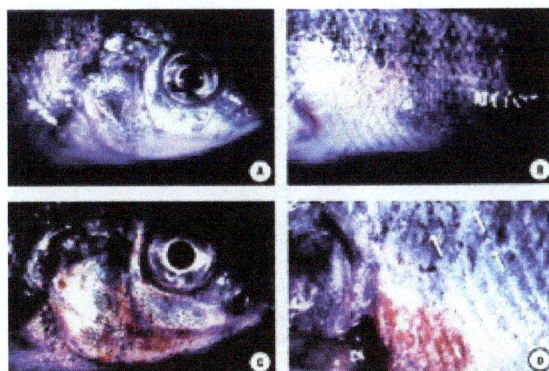

Skin ulcers in tilapia exposed to Pfiesteria shumwayae

Immune System

Immune organs vary by type of fish. In the jawless fish (lampreys and hagfish), true lymphoid organs are absent. These fish rely on regions of lymphoid tissue within other organs to produce immune cells. For example, erythrocytes, macrophages and plasma cells are produced in the anterior kidney (or pronephros) and some areas of the gut (where granulocytes mature.) They resemble primitive bone marrow in hagfish. Cartilaginous fish (sharks and rays) have a more advanced immune system. They have three specialized organs that are unique to chondrichthyes; the epigonal organs (lymphoid tissue similar to mammalian bone) that surround the gonads, the Leydig's organ within the walls of their esophagus, and a spiral valve in their intestine. These organs house typical immune cells (granulocytes, lymphocytes and plasma cells). They also possess an identifiable thymus and a well-developed spleen (their most important immune organ) where various lymphocytes, plasma cells and macrophages develop and are stored. Chondrostean fish (sturgeons, paddlefish and bichirs) possess a major site for the production of granulocytes within a mass that is associated with the meninges (membranes surrounding the central nervous system.) Their heart is frequently covered with tissue that contains lymphocytes, reticular cells and a small number of macrophages. The chondrostean kidney is an important hemopoietic organ; where erythrocytes, granulocytes, lymphocytes and macrophages develop.

Like chondrostean fish, the major immune tissues of bony fish (or teleostei) include the kidney (especially the anterior kidney), which houses many different immune cells. In addition, teleost fish possess a thymus, spleen and scattered immune areas within mucosal tissues (e.g. in the skin, gills, gut and gonads). Much like the mammalian immune system, teleost erythrocytes, neutrophils and granulocytes are believed to reside in the spleen whereas lymphocytes are the major cell type found in the thymus. In 2006, a lymphatic system similar to that in mammals was described in one species of teleost fish, the zebrafish. Although not confirmed as yet, this system presumably will be where naive (unstimulated) T cells accumulate while waiting to encounter an antigen.

Spreading Disease and Parasites

The capture, transportation and culture of bait fish can spread damaging organisms between ecosystems, endangering them. In 2007, several American states, including Michigan, enacted regulations designed to slow the spread of fish diseases, including viral hemorrhagic septicemia, by bait fish. Because of the risk of transmitting *Myxobolus cerebralis* (whirling disease), trout and salmon should not be used as bait. Anglers may increase the possibility of contamination by emptying bait buckets into fishing venues and collecting or using bait improperly. The transportation of fish from one location to another can break the law and cause the introduction of fish and parasites alien to the ecosystem.

Eating Raw Fish

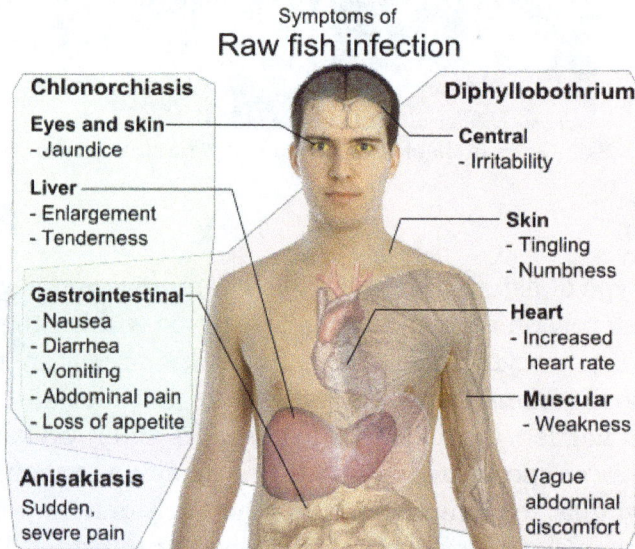

Differential symptoms of parasite infection by raw fish: Clonorchis sinensis (a trematode/fluke), Anisakis (a nematode/roundworm) and Diphyllobothrium a (cestode/tapeworm), all have gastrointestinal, but otherwise distinct, symptoms.

Though not a health concern in thoroughly cooked fish, parasites are a concern when human consumers eat raw or lightly preserved fish such as sashimi, sushi, ceviche, and gravlax. The popularity of such raw fish dishes makes it important for consumers to be aware of this risk. Raw fish should be frozen to an internal temperature of −20 °C (−4 °F) for at least 7 days to kill parasites. It is important to be aware that home freezers may not be cold enough to kill parasites.

Traditionally, fish that live all or part of their lives in fresh water were considered unsuitable for sashimi due to the possibility of parasites. Parasitic infections from freshwater fish are a serious problem in some parts of the world, particularly Southeast Asia. Fish that spend part of their life cycle in salt water, like salmon, can also be a problem. A study in Seattle, Washington showed that 100% of wild salmon had roundworm larvae capable of infecting people. In the same study farm raised salmon did not have any roundworm larvae.

Parasite infection by raw fish is rare in the developed world (fewer than 40 cases per year in the U.S.), and involves mainly three kinds of parasites: Clonorchis sinensis (a trematode/fluke), Anisakis (a nematode/roundworm) and Diphyllobothrium (a cestode/tapeworm). Infection by the fish tapeworm *Diphyllobothrium latum* is seen in countries where people eat raw or undercooked fish, such as some countries in Asia, Eastern Europe, Scandinavia, Africa, and North and South America. Infection risk of anisakis is particularly higher in fishes which may live in a river such as salmon (*shake*) in Salmonidae, mackerel (*saba*). Such parasite infections can generally be avoided by boiling, burning, preserving in salt or vinegar, or freezing overnight. Even Japanese people never eat raw salmon or ikura (salmon roe), and even if they seem raw, these foods are not raw but are frozen overnight to prevent infections from parasites, particularly anisakis.

Below are some life cycles of fish parasites that can infect humans:

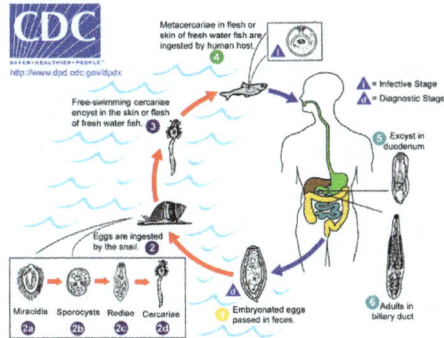

Life cycle of the liver fluke *Clonorchis sinensis*

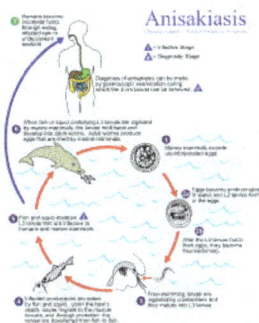

Life cycle of the parasitic *Anisakis* worm

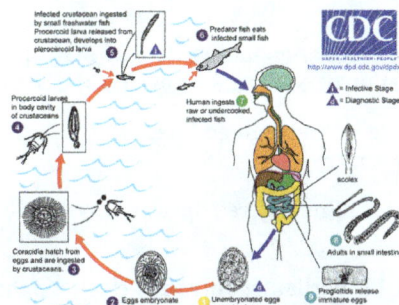

Life cycle of the fish tapeworm *Diphyllobothrium latum*

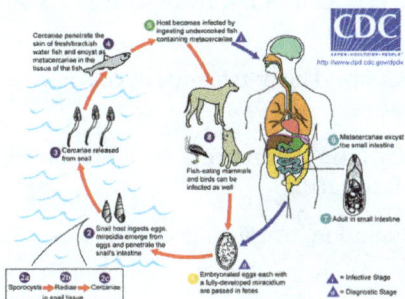

Life cycle of the digenean *Metagonimus*, an intestinal fluke

References

- Clucas, I.; Teutscher, F., eds. (1999). FAO/DFID Expert Consultation on Bycatch Utilization in Tropical Fisheries. Beijing (China), 21–28 Sep 1998. University of Greenwich, NRI. p. 333. ISBN 0-85954-504-0.

- Exell A, Burgess PH, Bailey MT. A-Z of Tropical Fish Diseases and Health Problems. New York, N.Y: Howell Book House. ISBN 1-58245-049-8.

- Fairfield, T (2000). A commonsense guide to fish health. Woodbury, N.Y: Barron's Educational Series. ISBN 0-7641-1338-0.

- Moyle, PB and Cech, JJ (2004) Fishes, An Introduction to Ichthyology. 5th Ed, Benjamin Cummings. ISBN 978-0-13-100847-2

- Woo PTK (1995) Fish Diseases and Disorders: Volume 1: Protozoan and Metazoan Infections Cabi Series. ISBN 9780851988238.

- Woo PTK (2011) Fish Diseases and Disorders: Volume 2: Non-Infectious Disorders Cabi Series. ISBN 9781845935535.

- Woo PTK (2011) Fish Diseases and Disorders: Volume 3: Viral, Bacterial and Fungal Infections Cabi Series. ISBN 9781845935542.

- "Welfare Aspects of Animal Stunning and Killing Methods" (PDF). European Food Safety Authority. 15 Jun 2004. Retrieved 21 January 2015.

- Yue, Stephanie. "An HSUS Report: The Welfare of Farmed Fish at Slaughter" (PDF). Humane Society of the United States. Retrieved 21 January 2015.

- Mood, A.; P. Brooke. "Estimating the number of fish caught in global fishing each year" (PDF). fishcount.org.uk. Retrieved 8 April 2014.

- Mood, A.; P. Brooke. "Estimating the number of farmed fish killed in global aquaculture each year" (PDF). fishcount.org.uk. Retrieved 8 April 2014.

- Farm Animal Welfare Council. "Report on the welfare of farmed fish - Recommendations: trout". Retrieved 13 February 2014.

- Pollard D and Smith A (2009). "Carcharias taurus". IUCN Red List of Threatened Species. Version 2011.2. International Union for Conservation of Nature. Retrieved 9 March 2013.

- Convention on International Trade in Endangered Species of Wild Fauna and Flora (2013). "Appendices I, II and III". Retrieved 13 September 2013.3

- Convention on International Trade in Endangered Species of Wild Fauna and Flora (2013). "The CITES Appendices". Retrieved 13 September 2013.

- Banks, M. (25 September 2012). "EU parliament urged to close 'loopholes' in shark fin legislation". The Parliament.com. Retrieved 13 September 2013.

- Kari, Shannon (30 November 2012). "Ontario court overturns Toronto's shark fin ban". The Globe and Mail. Retrieved 19 January 2013.

- Raloff, Janet. "No Way to Make Soup—Thirty-two tons of contraband shark fins seized on the high seas". Science News. Retrieved 25 March 2012.

- Rosenthal, Elisabeth (22 February 2012). "New York May Ban Shark Fin Sales, Following Other States". The New York Times. Retrieved 25 March 2012.

- Lu Hui, ed. (16 August 2006). "Pollution, overfishing destroy East China Sea fishery". Xinhua on GOV.cn. Retrieved 2012-05-01.

- Scales, Helen (29 March 2007). "Shark Declines Threaten Shellfish Stocks, Study Says". National Geographic News. Retrieved 2012-05-01.

6

Effect of Climate Change on Fisheries

The effects of global warming and pollution have affected the fishing industry as well. Climate change and the acidification of oceans are strikingly changing aquatic ecosystems. The rise in sea-levels directly affect communities and groups that practice fishing and aquaculture. This chapter focuses on the effects of climate change, its impact on the oceans and other water bodies and also on fish stocks.

Fisheries and Climate Change

Fishing with a lift net in Bangladesh. Coastal fishing communities in Bangladesh are vulnerable to flooding from sea-level rises.

Rising ocean temperatures and ocean acidification are radically altering aquatic ecosystems. Climate change is modifying fish distribution and the productivity of marine and freshwater species. This has impacts on the sustainability of fisheries and aquaculture, on the livelihoods of the communities that depend on fisheries, and on the ability of the oceans to capture and store carbon (biological pump). The effect of sea level rise means that coastal fishing communities are in the front line of climate change, while changing rainfall patterns and water use impact on inland (freshwater) fisheries and aquaculture.

Role of Oceans

Island with fringing reef in the Maldives. Coral reefs are dying around the world.

Oceans and coastal ecosystems play an important role in the global carbon cycle and have removed about 25% of the carbon dioxide emitted by human activities between 2000 and 2007 and about half the anthropogenic CO_2 released since the start of the Industrial Revolution. Rising ocean temperatures and ocean acidification means that the capacity of the ocean carbon sink will gradually get weaker, giving rise to global concerns expressed in the Monaco and Manado Declarations. Healthy ocean ecosystems are essential for the mitigation of climate change. Coral reefs provide habitat for millions of fish species and with climate change it can provoke these reefs to die.

Impact on Fish Production

The rising ocean acidity makes it more difficult for marine organisms such as shrimps, oysters, or corals to form their shells – a process known as calcification. Many important animals, such as zooplankton, that forms the base of the marine food chain have calcium shells. Thus the entire marine food web is being altered – there are 'cracks in the food chain'. As a result, the distribution, productivity, and species composition of global fish production is changing, generating complex and inter-related impacts on oceans, estuaries, coral reefs, mangroves and sea grass beds that provide habitats and nursery areas for fish. Changing rainfall patterns and water scarcity is impacting on river and lake fisheries and aquaculture production. After the ice age about 200,000 years ago, the global air temperature has risen 3 degrees, leading to an increase in sea temperatures.

Impact on Fishing Communities

Coastal and fishing populations and countries dependent on fisheries are particularly vulnerable to climate change. Low-lying countries such as the Maldives and Tuvalu are particularly vulnerable and entire communities may become the first climate refugees. Fishing communities in Bangladesh are subject not only to sea-level rise, but also flooding and increased typhoons. Fishing communities along the Mekong river produce over 1 million tons of basa fish annually and livelihoods and fish production will suffer from saltwater intrusion resulting from rising sea level and dams.

Fisheries and aquaculture contribute significantly to food security and livelihoods. Fish provides essential nutrition for 3 billion people and at least 50% of animal protein and minerals to 400 million people from the poorest countries. Over 500 million people in developing countries depend, directly or indirectly, on fisheries and aquaculture for their livelihoods - aquaculture is the world's fastest growing food production system, growing at 7% annually and fish products are among the most widely traded foods, with more than 37% (by volume) of world production traded internationally.

Fisherman landing his catch, Seychelles

Adaptation and Mitigation

The impacts of climate change can be addressed through adaptation and mitigation. The costs and benefits of adaptation are essentially local or national, while the costs of mitigation are essentially national whereas the benefits are global. Some activities generate both mitigation and adaptation benefits, for example, the restoration of mangrove forests can protect shorelines from erosion and provide breeding grounds for fish while also sequestering carbon.

Adaptation

Several international agencies, including the World Bank and the Food and Agriculture Organization have programs to help countries and communities adapt to global warming, for example by developing policies to improve the resilience of natural resources, through assessments of risk and vulnerability, by increasing awareness of climate change impacts and strengthening key institutions, such as for weather forecasting and early warning systems. The World Development Report 2010 - Development and Climate Change, Chapter 3 shows that reducing overcapacity in fishing fleets and rebuilding fish stocks can both improve resilience to climate change and increase economic returns from marine capture fisheries by US$50 billion per year, while also reducing GHG emissions by fishing fleets. Consequently, removal of subsidies on fuel for fishing can have a double benefit by reducing emissions and overfishing.

Investment in sustainable aquaculture can buffer water use in agriculture while producing food and diversifying economic activities. Algal biofuels also show potential as algae can produce 15-

300 times more oil per acre than conventional crops, such as rapeseed, soybeans, or jatropha and marine algae do not require scarce freshwater. Programs such as the GEF-funded Coral Reef Targeted Research provide advice on building resilience and conserving coral reef ecosystems, while six Pacific countries recently gave a formal undertaking to protect the reefs in a biodiversity hotspot – the Coral Triangle.

Mitigation

The White Cliffs of Dover

The oceans have removed 50% of the anthropogenic CO_2, so the oceans have absorbed much of the impact of climate change. The famous White Cliffs of Dover illustrate how the ocean captures and buries carbon. These limestone cliffs are formed from the skeletons of marine plankton called coccoliths. Similarly, petroleum formation is attributed largely to marine and aquatic plankton further illustrating the key role of the oceans in carbon sequestration.

Exactly how the oceans capture and bury CO_2 is the subject of intense research by scientists world-wide, such as the Carboocean Project. The current level of GHG emissions means that ocean acidity will continue to increase and aquatic ecosystems will continue to degrade and change. There are feedback mechanisms involved here. For example, warmer waters can absorb less CO_2, so as ocean temperatures rise some dissolved CO_2 will be released back into the atmosphere. Warming also reduces nutrient levels in the mesopelagic zone (about 200 to 1000 m deep). This in turn limits the growth of diatoms in favour of smaller phytoplankton that are poorer biological pumps of carbon. This inhibits the ability of the ocean ecosystems to sequester carbon as the oceans warm. What is clear, is that healthy ocean and coastal ecosystems are necessary to continue the vital role of the ocean carbon sinks, as indicated, for example, by the Blue Carbon assessment prepared by UNEP and the coastal carbon sinks report of IUCN and growing evidence of the role of fish biomass in the transport of carbon from surface waters to the deep ocean.

While the various carbon finance instruments include restoration of forests (REDD) and producing clean energy (emissions trading), few address the need to finance healthy ocean and aquatic ecosystems although these are essential for continued uptake of CO_2 and GHGs. The scientific basis for ocean fertilization – to produce more phytoplankton to increase the uptake of CO_2 – has been challenged, and proposals for burial of CO_2 in the deep ocean have come under criticism from environmentalists. The debate on these issues has underlined the need to increase scientific understanding of how the ocean sequesters carbon.

Over-fishing

Although there is a decline of fisheries due to climate change, another impact for this decrease is due to over-fishing. Studies show that the state of the ocean is causing fisheries to collapse, and in areas where fisheries have not yet collapsed, the amount of over fishing that is done is having a significant impact on the industry. Over-fishing is due to having access to the open sea, it makes it very easy for people to over fish, even if it is just for fun. There is also a high demand for sea food by fishermen, as well modern technology that has increased the amount of fish caught during each trip.

If there was a specific amount of fish that people were allowed to catch then this could very well solve the problem of over fishing. This type of limit system is in place in a few countries including New Zealand, Norway, Canada, and the United States. In these countries the limit system has successfully helped in fishing industries. These types of limit systems are called Individual fishing quota. This means that the areas where this quota exist, the government has legal entity over it and in these boundaries they are entitled to utilize their ocean resources as they wish.

References

- Klyashtorin LB (2001) Climate change and long-term fluctuations of commercial catches: the possibility of forecasting Technical paper 410, FAO fisheries, Rome. ISBN 978-92-5-104695-1.

7

Population Management of Fisheries

The term population dynamics in fisheries refers to the growth or fall in the population of aquatic organisms. It forms a basis for understanding the issues of habitat destruction, optimal harvesting rates and predation. Fisheries depend on effective analysis of data gathered through national and international efforts. This chapter deals exclusively with the tools and theories used in measuring population dynamics and understanding viable and sustainable fishing techniques.

Population Dynamics of Fisheries

Predator bluefin trevally sizing up schooling anchovies, in the Maldives

A fishery is an area with an associated fish or aquatic population which is harvested for its commercial or recreational value. Fisheries can be wild or farmed. Population dynamics describes the ways in which a given population grows and shrinks over time, as controlled by birth, death, and migration. It is the basis for understanding changing fishery patterns and issues such as habitat destruction, predation and optimal harvesting rates. The population dynamics of fisheries is used by fisheries scientists to determine sustainable yields.

The basic accounting relation for population dynamics is the BIDE (Birth, Immigration, Death, Emigration) model, shown as:

$$N_1 = N_0 + B - D + I - E$$

where N_1 is the number of individuals at time 1, N_0 is the number of individuals at time 0, B is the number of individuals born, D the number that died, I the number that immigrated, and E

the number that emigrated between time 0 and time 1. While immigration and emigration can be present in wild fisheries, they are usually not measured.

A fishery population is affected by three dynamic rate functions:

- Birth rate or recruitment. Recruitment means reaching a certain size or reproductive stage. With fisheries, recruitment usually refers to the age a fish can be caught and counted in nets.

- Growth rate. This measures the growth of individuals in size and length. This is important in fisheries where the population is often measured in terms of biomass.

- Mortality. This includes harvest mortality and natural mortality. Natural mortality includes non-human predation, disease and old age.

If these rates are measured over different time intervals, the harvestable surplus of a fishery can be determined. The harvestable surplus is the number of individuals that can be harvested from the population without affecting long term stability (average population size). The harvest within the harvestable surplus is called compensatory mortality, where the harvest deaths are substituting for the deaths that would otherwise occur naturally. Harvest beyond that is additive mortality, harvest in addition to all the animals that would have died naturally.

Care is needed when applying population dynamics to real world fisheries. Over-simplistic modelling of fisheries has resulted in the collapse of key stocks.

History

The first principle of population dynamics is widely regarded as the exponential law of Malthus, as modelled by the Malthusian growth model. The early period was dominated by demographic studies such as the work of Benjamin Gompertz and Pierre François Verhulst in the early 19th century, who refined and adjusted the Malthusian demographic model. A more general model formulation was proposed by F.J. Richards in 1959, by which the models of Gompertz, Verhulst and also Ludwig von Bertalanffy are covered as special cases of the general formulation.

Population Size

The population size (usually denoted by N) is the number of individual organisms in a population.

The effective population size (N_e) was defined by Sewall Wright, who wrote two landmark papers on it (Wright 1931, 1938). He defined it as "the number of breeding individuals in an idealized population that would show the same amount of dispersion of allele frequencies under random genetic drift or the same amount of inbreeding as the population under consideration". It is a basic parameter in many models in population genetics. N_e is usually less than N (the absolute population size).

Small population size results in increased genetic drift. Population bottlenecks are when population size reduces for a short period of time.

Overpopulation may indicate any case in which the population of any species of animal may exceed the carrying capacity of its ecological niche.

Virtual Population Analysis

Virtual population analysis (VPA) is a cohort modeling technique commonly used in fisheries science for reconstructing historical fish numbers at age using information on death of individuals each year. This death is usually partitioned into catch by fisheries and natural mortality. VPA is virtual in the sense that the population size is not observed or measured directly but is inferred or back-calculated to have been a certain size in the past in order to support the observed fish catches and an assumed death rate owing to non-fishery related causes.

Minimum Viable Population

The minimum viable population (MVP) is a lower bound on the population of a species, such that it can survive in the wild. More specifically MVP is the smallest possible size at which a biological population can exist without facing extinction from natural disasters or demographic, environmental, or genetic stochasticity. The term "population" refers to the population of a species in the wild.

As a reference standard, MVP is usually given with a population survival probability of somewhere between ninety and ninety-five percent and calculated for between one hundred and one thousand years into the future.

The MVP can be calculated using computer simulations known as population viability analyses (PVA), where populations are modelled and future population dynamics are projected.

Maximum Sustainable Yield

In population ecology and economics, the maximum sustainable yield or MSY is, theoretically, the largest catch that can be taken from a fishery stock over an indefinite period. Under the assumption of logistic growth, the MSY will be exactly at half the carrying capacity of a species, as this is the stage at when population growth is highest. The maximum sustainable yield is usually higher than the optimum sustainable yield.

This logistic model of growth is produced by a population introduced to a new habitat or with very poor numbers going through a lag phase of slow growth at first. Once it reaches a foothold population it will go through a rapid growth rate that will start to level off once the species approaches carrying capacity. The idea of maximum sustained yield is to decrease population density to the point of highest growth rate possible. This changes the number of the population, but the new number can be maintained indefinitely, ideally.

MSY is extensively used for fisheries management. Unlike the logistic (Schaefer) model, MSY in most modern fisheries models occurs at around 30% of the unexploited population size. This fraction differs among populations depending on the life history of the species and the age-specific selectivity of the fishing method.

However, the approach has been widely criticized as ignoring several key factors involved in fisheries management and has led to the devastating collapse of many fisheries. As a simple calculation, it ignores the size and age of the animal being taken, its reproductive status, and it focuses solely on the species in question, ignoring the damage to the ecosystem caused by the designated level of exploitation and the issue of bycatch. Among conservation biologists it is widely regarded as dangerous and misused.

Recruitment

Recruitment is the number of new young fish that enter a population in a given year. The size of fish populations can fluctuate by orders of magnitude over time, and five to 10-fold variations in abundance are usual. This variability applies across time spans ranging from a year to hundreds of years. Year to year fluctuations in the abundance of short lived forage fish can be nearly as great as the fluctuations that occur over decades or centuries. This suggests that fluctuations in reproductive and recruitment success are prime factors behind fluctuations in abundance. Annual fluctuations often seem random, and recruitment success often has a poor relationship to adult stock levels and fishing effort. This makes prediction difficult.

The recruitment problem is the problem of predicting the number of fish larvae in one season that will survive and become juvenile fish in the next season. It has been called "the central problem of fish population dynamics" and "the major problem in fisheries science". Fish produce huge volumes of larvae, but the volumes are very variable and mortality is high. This makes good predictions difficult.

According to Daniel Pauly, the definitive study was made in 1999 by Ransom Myers. Myers solved the problem "by assembling a large base of stock data and developing a complex mathematical model to sort it out. Out of that came the conclusion that a female in general produced three to five recruits per year for most fish."

Overfishing

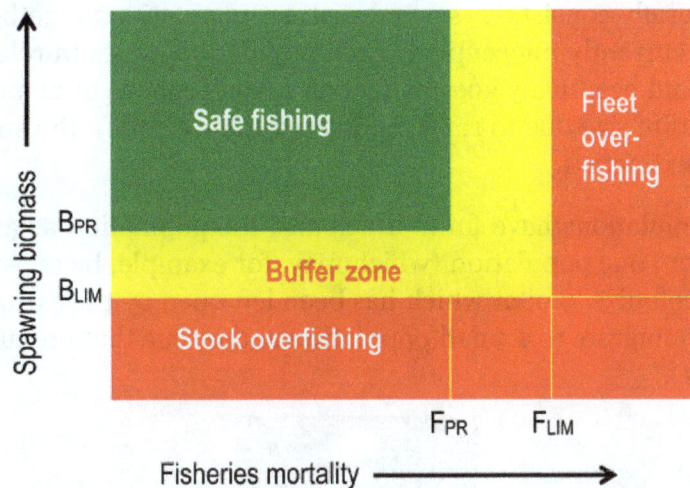

The Traffic Light colour convention, showing the concept of Harvest Control Rule (HCR), specifying when a rebuilding plan is mandatory in terms of precautionary and limit reference points for spawning biomass and fishing mortality rate.

The notion of overfishing hinges on what is meant by an acceptable level of fishing.

A current operational model used by some fisheries for predicting acceptable levels is the Harvest Control Rule (HCR). This formalizes and summarizes a management strategy which can actively adapt to subsequent feedback. The HCR is a variable over which the management has some direct control and describes how the harvest is intended to be controlled by management in relation to the state of some indicator of stock status. For example, a harvest control rule can describe the various values of fishing mortality which will be aimed at for various values of the stock abundance. Constant catch and constant fishing mortality are two types of simple harvest control rules.

- Biological overfishing occurs when fishing mortality has reached a level where the stock biomass has negative marginal growth (slowing down biomass growth), as indicated by the red area in the figure. Fish are being taken out of the water so quickly that the replenishment of stock by breeding slows down. If the replenishment continues to slow down for long enough, replenishment will go into reverse and the population will decrease.

- Economic or bioeconomic overfishing additionally considers the cost of fishing and defines overfishing as a situation of negative marginal growth of resource rent. Fish are being taken out of the water so quickly that the growth in the profitability of fishing slows down. If this continues for long enough, profitability will decrease.

Metapopulation

A metapopulation is a group of spatially separated populations of the same species which interact at some level. The term was coined by Richard Levins in 1969. The idea has been most broadly applied to species in naturally or artificially fragmented habitats. In Levins' own words, it consists of "a population of populations".

A metapopulation generally consists of several distinct populations together with areas of suitable habitat which are currently unoccupied. Each population cycles in relative independence of the other populations and eventually goes extinct as a consequence of demographic stochasticity (fluctuations in population size due to random demographic events); the smaller the population, the more prone it is to extinction.

Although individual populations have finite life-spans, the population as a whole is often stable because immigrants from one population (which may, for example, be experiencing a population boom) are likely to re-colonize habitat which has been left open by the extinction of another population. They may also emigrate to a small population and rescue that population from extinction (called the *rescue effect*).

Malthus

BENJAMIN GOMPERTZ.

Gompertz

P. F. VERHULST.

Verhulst

Age Class Structure

Age can be determined by counting growth rings in fish scales, otoliths, cross-sections of fin spines for species with thick spines such as triggerfish, or teeth for a few species. Each method has its merits and drawbacks. Fish scales are easiest to obtain, but may be unreliable if scales have fallen off of the fish and new ones grown in their places. Fin spines may be unreliable for the same reason, and most fish do not have spines of sufficient thickness for clear rings to be visible. Otoliths will have stayed with the fish throughout its life history, but obtaining them requires killing the fish. Also, otoliths often require more preparation before ageing can occur.

An age class structure with gaps in it, for instance a regular bell curve for the population of 1-5 year-old fish, excepting a very low population for the 3-year-olds, implies a bad spawning year 3 years ago in that species.

Often fish in younger age class structures have very low numbers because they were small enough to slip through the sampling nets, and may in fact have a very healthy population.

Population Cycle

A population cycle occurs where populations rise and fall over a predictable period of time. There are some species where population numbers have reasonably predictable patterns of change although the full reasons for population cycles is one of the major unsolved ecological problems. There are a number of factors which influence population change such as availability of food, predators, diseases and climate.

Trophic Cascades

Trophic cascades occur when predators in a food chain suppress the abundance of their prey, thereby releasing the next lower trophic level from predation (or herbivory if the intermediate trophic level is an herbivore). For example, if the abundance of large piscivorous fish is increased in a lake, the abundance of their prey, zooplanktivorous fish, should decrease, large zooplankton abundance should increase, and phytoplankton biomass should decrease. This theory has stimulated new research in many areas of ecology. Trophic cascades may also be important for understanding the effects of removing top predators from food webs, as humans have done in many places through hunting and fishing activities.

Classic examples

1. In lakes, piscivorous fish can dramatically reduce populations of zooplanktivorous fish, zooplanktivorous fish can dramatically alter freshwater zooplankton communities, and zooplankton grazing can in turn have large impacts on phytoplankton communities. Removal of piscivorous fish can change lake water from clear to green by allowing phytoplankton to flourish.

2. In the Eel River, in Northern California, fish (steelhead and roach) consume fish larvae and predatory insects. These smaller predators prey on midge larvae, which feed on algae. Removal of the larger fish increases the abundance of algae.

3. In Pacific kelp forests, sea otters feed on sea urchins. In areas where sea otters have been hunted to extinction, sea urchins increase in abundance and decimate kelp

A recent theory, the mesopredator release hypothesis, states that the decline of top predators in an ecosystem results in increased populations of medium-sized predators (mesopredators).

Basic Models

*   The classic population equilibrium model is Verhulst's 1838 growth model:

    $$\frac{dN}{dt} = rN\left(1 - \frac{N}{K}\right)$$

    where $N(t)$ represents number of individuals at time t, r the intrinsic growth rate and K is the carrying capacity, or the maximum number of individuals that the environment can support.

*   The individual growth model, published by von Bertalanffy in 1934, can be used to model the rate at which fish grow. It exists in a number of versions, but in its simplest form it is expressed as a differential equation of length (L) over time (t):

    $$L'(t) = r_B\left(L_\infty - L(t)\right)$$

    where r_B is the von Bertalanffy growth rate and L_∞ the ultimate length of the individual.

*   Schaefer published a fishery equilibrium model based on the Verhulst model with an assumption of a bi-linear catch equation, often referred to as the Schaefer short-term catch equation:

$$H(E, X) = qEX$$

where the variables are; H, referring to catch (harvest) over a given period of time (e.g. a year); E, the fishing effort over the given period; X, the fish stock biomass at the beginning of the period (or the average biomass), and the parameter q represents the catchability of the stock.

Assuming the catch to equal the net natural growth in the population over the same period ($\dot{X} = 0$),), the equilibrium catch is a function of the long term fishing effort E:

$$H(E) = qKE(1 - \frac{qE}{r})$$

r and K being biological parameters representing intrinsic growth rate and natural equilibrium biomass respectively.

- The Baranov catch equation of 1918 is perhaps the most used equation in fisheries modelling. It gives the catch in numbers as a function of initial population abundance N_o and fishing F and natural mortality M:

$$C = \frac{F}{F + M}(1 - e^{-(F+M)T})N_0$$

where T is the time period and is usually left out (i.e. $T=1$ is assumed). The equation assumes that fishing and natural mortality occur simultaneously and thus "compete" with each other. The first term expresses the proportion of deaths that are caused by fishing, and the second and third term the total number of deaths.

- The Ricker model is a classic discrete population model which gives the expected number (or density) of individuals N_{t+1} in generation $t + 1$ as a function of the number of individuals in the previous generation,

$$N_{t+1} = N_t e^{r(1 - \frac{N_t}{k})}$$

Here r is interpreted as an intrinsic growth rate and k as the carrying capacity of the environment. The Ricker model was introduced in the context of the fisheries by Ricker (1954).

- The Beverton–Holt model, introduced in the context of fisheries in 1957, is a classic discrete-time population model which gives the expected number n_{t+1} (or density) of individuals in generation $t + 1$ as a function of the number of individuals in the previous generation,

$$n_{t+1} = \frac{R_0 n_t}{1 + n_t / M}.$$

Here R_0 is interpreted as the proliferation rate per generation and $K = (R_0 - 1) M$ is the carrying capacity of the environment.

- Nurgaliev's law says

$$\frac{dN}{dt} = aN - bN$$

where N is the size of a population, a is a half of the average probability of a birth of a male (the same for females) of a potential arbitrary parents pair within a year, and b is an average probability of a death of a fish within a year.

Predator-prey Equations

The classic predator-prey equations are a pair of first order, non-linear, differential equations used to describe the dynamics of biological systems in which two species interact, one a predator and one its prey. They were proposed independently by Alfred J. Lotka in 1925 and Vito Volterra in 1926.

An extension to these are the competitive Lotka-Volterra equations, which provide a simple model of the population dynamics of species competing for some common resource.

In the 1930s Alexander Nicholson and Victor Bailey developed a model to describe the population dynamics of a coupled predator-prey system. The model assumes that predators search for prey at random, and that both predators and prey are assumed to be distributed in a non-contiguous ("clumped") fashion in the environment.

In the late 1980s, a credible, simple alternative to the Lotka-Volterra predator-prey model (and its common prey dependent generalizations) emerged, the ratio dependent or Arditi-Ginzburg model. The two are the extremes of the spectrum of predator interference models. According to the authors of the alternative view, the data show that true interactions in nature are so far from the Lotka-Volterra extreme on the interference spectrum that the model can simply be discounted as wrong. They are much closer to the ratio dependent extreme, so if a simple model is needed one can use the Arditi-Ginzburg model as the first approximation.

Virtual Population Analysis

Virtual population analysis (VPA) is a cohort modeling technique commonly used in fisheries science for reconstructing historical fish numbers at age using information on death of individuals each year. This death is usually partitioned into catch by fisheries and natural mortality. VPA is virtual in the sense that the population size is not observed or measured directly but is inferred or back-calculated to have been a certain size in the past in order to support the observed fish catches and an assumed death rate owing to non-fishery related causes.

Virtual population analysis was introduced in fish stock assessment by Gulland in 1965 based on older work. The technique of cohort reconstruction in fish populations has been attributed to several different workers including Professor Baranov from Russia in 1918 for his development of the continuous catch equation, Professor Fry from Canada in 1949 and Drs. Beverton and Holt from the UK in 1957. Because cohort reconstruction is essentially an accounting exercise it was likely independently conceived many times.

Several different software implementations of cohort reconstruction for fish populations exist including ADAPT, which is often used in Canada and the USA, and XSA which is commonly used in Europe. The back-calculations in these implementations work the same way but they differ in the statistical methods used for "tuning" to indices of population size.

Minimum Viable Population

Minimum viable population (MVP) is a lower bound on the population of a species, such that it can survive in the wild. This term is used in the fields of biology, ecology, and conservation biology. More specifically, MVP is the smallest possible size at which a biological population can exist without facing extinction from natural disasters or demographic, environmental, or genetic stochasticity. The term "population" rarely refers to an entire species. For example, the undomesticated dromedary camel is extinct in its natural wild habitat; but there is a domestic population in captivity and an additional feral population in Australia. Two groups of house cats in separate houses which are not allowed outdoors are also technically distinct populations. Typically, however, MVP is used to refer solely to a wild population, such as the red wolf.

Estimation

Minimum viable population is usually estimated as the population size necessary to ensure between 90 and 95 percent probability of survival between 100 and 1,000 years into the future. The MVP can be estimated using computer simulations for population viability analyses (PVA). PVA models populations using demographic and environmental information to project future population dynamics. The probability assigned to a PVA is arrived at after repeating the environmental simulation thousands of times.

For example, for a theoretical simulation of a population of 50 giant pandas in which the simulated population goes completely extinct, 30 out of 100 stochastic simulations projected 100 years into the future are not viable. Causes of extinction in the simulation may include inbreeding depression, natural disaster, or climate change. Extinction occurring in 30 out of 100 runs would give a survival probability of 70%. In contrast, in the same simulation with a starting population of 60 pandas, the panda population may only become extinct in four of the hundred runs, resulting in a survival probability of 96%. In this case the minimum viable population that satisfies the 90- to 95% probability for survival is between 50 and 60 pandas. (These figures have been invented for the purpose of this example.)

Mvp and Extinction

MVP does not take human intervention into account. Thus, it is useful for conservation managers and environmentalists; a population may be increased above the MVP using a captive breeding program, or by bringing other members of the species in from other reserves.

There is naturally some debate on the accuracy of PVAs, since a wide variety of assumptions generally are required for future forecasting; however, the important consideration is not absolute accuracy, but promulgation of the concept that each species indeed has an MVP, which at least can be approximated for the sake of conservation biology and Biodiversity Action Plans.

There is a marked trend for insularity, surviving genetic bottlenecks and r-strategy to allow far lower MVPs than average. Conversely, taxa easily affected by inbreeding depression – having high

MVPs – are often decidedly K-strategists, with low population densities while occurring over a wide range. An MVP of 500 to 1,000 has often been given as an average for terrestrial vertebrates when inbreeding or genetic variability is ignored. When inbreeding effects are included, estimates of MVP for many species are in the thousands. Based on a meta-analysis of reported values in the literature for many species, Traill *et al.* reported a median MVP of 4,169 individuals.

In 1912, the Laysan duck had an effective population size of 7 at most.

Population Uncertainty

Population uncertainty may be divided into four sources:

- Demographic stochasticity

- Environmental stochasticity

- Natural catastrophes

- Genetic stochasticity

References

- Kazan-Zelenodolsk; "'Law' of Two Hundred Billions in Context of Civil Society". In materials of Inter-regional scientific-practical conference The Civil Society: Ideas, Reality, Prospects, on April 27, 2006, p. 204-207. ISBN 5-8399-0153-9.

- Sparre, Per and Hart, Paul J B (2002) Handbook of Fish Biology and Fisheries, Chapter13: Choosing the best model for fisheries assessment. Blackwell Publishing. ISBN 0-632-06482-X

8

Laws Related to Fisheries

Understanding the complexities of thefishing industry and that disputes can arise in international waters, several laws and regulations are in place to allow judicious fishing while conserving ecosystems. This chapter provides knowledge about seafood safety regulations, individual fishing quotas while also tackling issues regarding the cultivation of genetically modified organisms (GMOs). Fisheries management is best understood in confluence with the major topics listed in the following chapter.

Fisheries Law

Fishery on Lake Tondano, Indonesia

Fisheries law is an emerging and specialized area of law. Fisheries law is the study and analysis of different fisheries management approaches such as catch shares e.g. Individual Transferable Quotas; TURFs; and others. The study of fisheries law is important in order to craft policy guidelines that maximize sustainability and legal enforcement. This specific legal area is rarely taught at law schools around the world, which leaves a vacuum of advocacy and research. Fisheries law also takes into account international treaties and industry norms in order to analyze fisheries management regulations. In addition, fisheries law includes access to justice for small-scale fisheries and coastal and aboriginal communities and labor issues such as child labor laws, employment law, and family law.

Another important area of research covered in fisheries law is seafood safety. Each country, or region, around the world has a varying degree of seafood safety standards and regulations. These regulations can contain a large diversity of fisheries management schemes including quota or catch share systems. It is important to study seafood safety regulations around the world in order to craft policy guidelines from countries who have implemented effective schemes. Also, this body of research can identify areas of improvement for countries who have not yet been able to master efficient and effective seafood safety regulations.

Fisheries law also includes the study of aquaculture laws and regulations. Aquaculture, also known as aquafarming, is the farming of aquatic organisms, such as fish and aquatic plants. This body of research also encompasses animal feed regulations and requirements. It is important to regulate what feed is consumed by fish in order to prevent risks to human health and safety.

Seafood Safety Regulations

U.S. Labeling of Genetically Engineered Salmon

On November 19, 2015, the U.S. Food and Drug Administration (FDA) approved AquaBounty Technologies' application to sell the AquAdvantage salmon to U.S. consumers. The possible introduction of genetically engineered salmon into the marketplace furthers discussion involving ethics, protection of the natural environment, international and domestic trade law, labeling practices, nutrition, and constitutional issues. As the FDA points out, nutrition labeling for raw produce (fruits and vegetables), fish, and genetically modified products, is voluntary. Under section 403(a)(1) of the *Food, Drug, and Cosmetic Act* ("FD&C Act"), a food is misbranded if its labeling is "false or misleading in any particular". Section 201(n) of the FD&C Act provides that labeling is misleading if it fails to reveal facts that are material in light of representations made or suggested in the labeling. In regards to the AquAdvantage salmon, the FDA has stated:

"Based on our assessments of food derived from the AquAdvantage Salmon, we have determined that the term "Atlantic salmon" is the appropriate common or usual name for such food within the meaning of section 403(i) of the FD&C Act because AquAdvantage Salmon meets FDA's regulatory standard for Atlantic salmon (Ref. 10) and the composition and basic nature of food from AquAdvantage Salmon does not significantly differ from its non-GE counterpart—non-GE farm-raised Atlantic salmon. In addition, we have determined that food derived from AquAdvantage Salmon is as safe and nutritious as food from other farm-raised Atlantic salmon. For these reasons, we have concluded that there is no material difference between food derived from AquAdvantage Salmon and food derived from other non-GE, farm-raised Atlantic salmon that is required to be disclosed in the labeling of food derived from AquAdvantage Salmon under the relevant provisions of the FD&C Act, as explained above."

Canadian Labeling of Genetically Engineered Salmon

In 2014, Canada was a top-five producer of GM crops in the world, and is one of the largest players on the global market. Health Canada and the Canadian Food Inspection Agency ("CFIA") carry joint responsibility for federally food labeling policies in Canada under the *Food and Drugs Act* ("F&D Act"). Under the F&D Act, GMOs are defined as "novel foods". A novel food is allowed to enter the Canadian marketplace only after it has passed an assessment undertaken by various stakeholders. Health Canada is responsible for deciding all health and safety labeling policies for food products, such as special dietary needs or GM products, and ensuring that the products are safe for consumption. The CFIA is responsible for the development of all non-health and safety food labeling regulations and policies and the enforcement of labeling legislation. Standards are set by the CFIA so that Canadian food labels are truthful and not misleading.

Subsection 5(1) of the F&D Act states that "no person shall label, package, treat, process, sell or advertise any food in a manner that is false, misleading or deceptive or is likely to create an erroneous impression regarding its character, value, quantity, composition, merit or safety". Food must be labelled only if there are changes in the food such as problematic allergens or a significant nutrient or compositional change. Health Canada has not released an official statement concerning the AquAdvantage salmon, but does state on its website: "after twelve years of reviewing the safety of novel foods, Health Canada is not aware of any published scientific evidence demonstrating that novel foods are any less safe than traditional foods. The regulatory framework put in place by the federal government ensures that new and modified foods can be safely introduced in to the Canadian diet".

Individual Fishing Quota

Individual fishing quotas (IFQs) also known as "individual transferable quotas" are one kind of *catch share*, a means by which many governments regulate fishing. The regulator sets a species-specific total allowable catch (TAC), typically by weight and for a given time period. A dedicated portion of the TAC, called quota shares, is then allocated to individuals. Quotas can typically be bought, sold and leased, a feature called transferability. As of 2008, 148 major fisheries (generally, a single species in a single fishing ground) around the world had adopted some variant of this approach, along with approximately 100 smaller fisheries in individual countries. Approximately 10% of the marine harvest was managed by ITQs as of 2008. The first countries to adopt individual fishing quotas were the Netherlands, Iceland and Canada in the late 1970s, and the most recent is the United States Scallop General Category IFQ Program in 2010. The first country to adopt individual transferable quotas as a national policy was New Zealand in 1986.

Command and Control Approaches

Historically, inshore and deep water fisheries were in common ownership where no one had a property right to the fish (i.e., owned them) until after they had been caught. Each boat faced the zero-sum game imperative of catching as many fish as possible, knowing that any fish they did not catch would likely be taken by another boat.

Commercial fishing evolved from subsistence fishing with no restrictions that would limit or direct the catch. The implicit assumption was that the ocean's bounty was so vast that restrictions were unnecessary. In the twentieth century, fisheries such as Atlantic cod and California sardines collapsed, and nations began to limit access to their fishing grounds by boats from other countries, while in parallel, international organizations began to certify that specific species were "threatened", "endangered", etc.

One early management technique was to define a "season" during which fishing was allowed. The length of the season attempted to reflect the current abundance of the fishery, with bigger populations supporting longer seasons. This turned fishing into a race, driving the industry to bigger, faster boats, which in turned caused regulators to repetitively shorten seasons, sometimes to only a few days per year. Landing all boats over an ever-shorter interval also led to glut/shortage market cycles with prices crashing when the boats came in. A secondary consequence was that boats sometimes embarked when the fishery was "open" regardless of weather or other safety concerns.

A Move to Privatization and Market Based Mechanisms

The implementation of ITQs or IFQs works in tandem with the privatization of common assets. This regulatory measure seeks to economically rationalise access to a common-pool resource. This type of management is based in the doctrine of natural resource economics. Notably the use of ITQs in environmental policy has been informed by the work of economists such as Jens Warming, H. Scott Gordon and Anthony Scott. It is theorised that the primary driver of over-fishing is the rule of capture externality. This is the idea that the fisher does not have a property right to the resource until point of capture, incentivising competitive behavior and overcapitalisation in the industry. It is theorized that without a long-term right to fish stocks, there is no incentive to conserve fish stocks for the future.

The use of ITQs in resource management dates back to the 1960s and was first seen in 'pollution quotas', which are now widely used to manage carbon emissions from power utilities. For both air and marine resources ITQs use a 'cap-and-trade' approach by setting typically annual limits on resource exploitation (TAC in fisheries) and then allowing trade of quotas between industry users. However, ITQ use in fisheries is fundamentally different from pollution quotas, since the latter regulates the byproduct of an industry, whereas fishery ITQ's regulate the actual output product of the fishing industry, and thus amount to exclusive industry participation rights.

The use of IFQs has often been related to broader processes within neoliberalism that tend to utilise markets as a regulatory tool. The rationale behind such neoliberal mechanisms situates itself in the belief that market mechanisms harness profit motive to more innovative and efficient environmental solutions than those devised and executed by states. Whilst such neoliberal regulation has often been posited as a move away from state governance, in the case of privatization the state is integral in the process of creating and maintaining property rights.

The use of neoliberal privatizing regimes has also often raised contradictions with the rights of indigenous communities. For example, the exclusion of the Maori in the initial allocation of fishing quota in New Zealand's quota management system led to a lengthy legal battle delaying development in national fisheries policy and resulting in a large settlement from the crown. There have also been similar legal battles regarding the allocation of fishing rights with the Mi'kmaq in Canada and the Saami in North Norway. Aboriginal fishing rights are said to pose a challenge to the authoritative claims of the state as the final arbitors in respect of access and participation in rights-based regimes.

Catch Shares

The term *catch share* has been used more recently to describe the range of programs similar to ITQs. Catch shares expanded the concept of daily catch limits to yearlong limits, allowed different fishers to have different limits based on various factors, and also limited the total catch.

Catch shares eliminate the "race to the fish" problem, because fishers are no longer restricted to short fishing seasons and can schedule their voyages as they choose. Boom/bust market cycles disappear, because fishing can continue throughout a typically many-month season. Some safety problems are reduced because there's no need to fish in hazardous conditions just because the fishery happens to be open.

A crucial element of catch share systems is how to distribute/allocate the shares and what rights come with them. The initial allocation can be granted or auctioned. Shares can be held permanently ("owned") or for a fixed period such as one year ("rented"). They can be salable and/or leasable or not, with or without limits. Each variation has advantages and disadvantages, which may vary given the culture of a given fishing community.

Initial Distribution

ITQs are typically initially allocated as grants according to the recent catch history of the fishery. Those with bigger catches generally get bigger quotas. The primary drawback is that individuals receive a valuable right at no cost. Grants are somewhat analogous to an "homestead", in which settlers who developed farms in the American wilderness eventually received title without payment to what had been public land. In some cases, less than 100% of the TAC becomes ITQs, with the remainder allocated to other management strategies.

The grant approach is inherently political, with attendant benefits and costs. For example, related industries such as fish processing and other non-participants may seek quota grants. Also, fishers are often excluded from receiving quota if they are not also boat owners, however boat owners who do not fish do receive quota, such as was the case in Alaskan IFQ distributions. The offshore pollock cooperative in the Pacific Northwest allocated initial quotas by mutual agreement and allows quota holders to sell their quotas only to the cooperative members.

Quota auctions recompense the public for access to fisheries. They are somewhat analogous to the spectrum auctions that the U.S. held to allocate highly valuable radio spectrum. These auctions raised 10s of billions of dollars for the public. Note however that the television industry did not have to pay for the necessary spectrum to switch from analog to digital broadcasting, which is more like quota grants for incumbent fishers.

Trading

ITQs can be resold to those who want to increase their presence in the fishery. Alternatively, quotas can be non-tradeable, meaning that if a fisher leaves the industry, the quota reverts to the government to retire or to grant/auction to another party.

Once distributed, quotas can be regranted/reauctioned periodically or held in perpetuity. Limiting the time period lowers the quota's value and its initial auction price/cost, but subsequent auctions create recurring revenues. At the same time, "privatizing" such a public resource reduces the remaining amount of public resources and can be thought of as "giving away our future". In the industry, rented quotas are often referred to as "dedicated access privileges" (DAP).

Another issue with tradability is that large enterprises may buy all the quotas, ending what may be a centuries-long tradition of small-scale operations. This may benefit the sellers (and the buyers and those who buy the fish) but can potentially cause large changes in the culture of fishing communities. Consolidation of quota accompanies every IFQ program, and typically works to phase out smaller, less profitable fishing operations in favor of larger, often corporate owned fleets who have better financing capabilities.

Some fisheries require quota holders to be participating fishermen to prevent absentee ownership and limit the quota that a captain can accumulate. In the Alaska halibut and black cod fisheries, only active fishers can buy quota, and new entrants may not sublease their quota. However, these measures have only served to mitigate outside speculation in IFQ's by non-fishers. A lack of regulatory policy or enforcement still results in the prevalence of "armchair fishermen" (those who own quota but do not materially participate in the fishery). Since IFQ's began in 1995, the commercial longline fleet has never exceeded these fisheries' TACs.

Other Characteristics

ITQs may have the effect of changing the criteria that fishers apply to their catch. Highgrading involves catching more fish than the quota allows and dumping specimens that are less valuable because of size, age or other criteria. Many of the discarded fish are already dead or quickly die, increasing fishing's impact on stocks.

Effectiveness

In 2008 a large scale study concluded that ITQs can help to prevent collapses and restore declining fisheries when compared to a data set including 11,000 fisheries of various management structures (some entirely unmanaged). While nearly a third of open-access fisheries have collapsed, catch share fisheries are only half as likely to fail. However, when compared to other modern fishery management schemes, IFQ managed fisheries exhibit no long term ecological advantages. A study of the 14 IFQ programs in the United States revealed that fish stocks are unaffected by these management schemes.

In 1995, the Alaskan halibut fishery converted to ITQs, after regulators cut the season from about four months down to two or three days. Today, due to the pre-allocation of catch that accompanies IFQ's, the season lasts nearly eight months and boats deliver fresh fish at a steadier pace. However, halibut stocks have been in continuous decline for over a decade, as poor stock assessments leading to overfishing has caused a substantial decline in biomass. Additionally, despite the increase in landings value, the number of quota holders have declined by 44%, as consolidation and quota pricing has served to prevent new entrants.

Not all fisheries have thrived under ITQs, in some cases experiencing reduced or static biomass levels, because of factors such as:

- TACs may be set at too high a level

- Migratory species may be overfished in parts of their habitat not covered by the TAC

- Habitats may incur damage

- Enforcement may be lax

In the United States

The Magnuson-Stevens Fishery Conservation and Management Act defines individual transferable quotas (ITQs) as permits to harvest specific quantities of fish of a particular species. Fisheries scientists decide the maximum annual harvest in a certain fishery, accounting for carrying capacity, regeneration rates and future values. This amount is called the total allowable catch (TAC). Under ITQs, participants in a fishery receive rights to a portion of the TAC without charge. Quotas can be fished, bought, sold, or leased. Twenty-eight U.S. fisheries have adopted ITQs as of 2008. Concerns about distributional impacts led to a moratorium on moving other fisheries into the program that lasted from 1996 to 2004.

Starting in January 2011, fishermen in California, Oregon and Washington will operate via tradeable catch shares. Fishers have been discarding bycatch that is not their target, typically killing the individuals. Catch shares allow trawlers to exchange bycatch with each other, benefiting both. Goals of the system include increased productivity, reduced waste, and higher revenues for fishers. More than a dozen other U.S. fisheries are now managed by catch shares. Fishery managers say that in Alaska, where catch shares have been in place for several years, fishermen are now getting higher prices for their catch.

Criticisms and Controversies

Private Control of Public Resource

IFQ's are usually initiated through the de facto privatization of an otherwise public resource: the fisheries. Initial recipients of quota receive windfall profits through the gifting of share ownership, while all future entrants are forced to purchase or lease the right harvest fish. Many have questioned both the ethical and economic repercussions of dedicating a secure, exclusive privilege to access this public resource. For example, in the US, during a presentation given to the Gulf Fishery Management Council, Fishery Manager Larry Abele stated that the present value of the Gulf Fishery IFQ Harvest amounted to $345,000,000 and this was given without requiring of any return to the public from IFQ holders.

Quota Consolidation

Virtually every IFQ program results in substantial consolidation of quota. For example, it is estimated that 8 companies control 80% of New Zealand's fisheries through quota acquisition, 4 companies control 77% of one Alaska crab fishery, and 7% of shareholders control 60% of the US Gulf Red Snapper quota. The consolidation results in job loss, reduced wages, and decreased entry opportunities into the fishery.

Leasing Practices

Many IFQ systems involve the temporary transfer of fishing rights, whereby the owner of quota leases the fishing rights to active fishermen in exchange for a fixed percentage of the landed value of fish. Since quota acquisition is often beyond the financial means of many fishermen, they are forced to sacrifice substantial portions of their income in order to lease fishing rights. For example, Bering Sea crab lease fees can be as high as 80% of the landed value of the crab, meaning that the active fishermen only retain 20% of the revenue, much which is needed to cover costs. In some fisheries, the majority of quota is leased to active fishermen, often by individuals who do not material participate in the fishery, but have been able to acquire shares. This makes quota acquisition even less likely for active fishermen, results in diversion of wealth away from fishing communities and into the hands of private investors, and can cause major financial strain on fishermen along with the economic contraction of fishing communities.

Economic Depression of Coastal Communities

The transition to IFQ management tends to cause considerable economic harm to coastal communities that are dependent on commercial fisheries. Although IFQ management systems are designed to enhance the economic performance of the fishing industry, this usually comes at the cost of coastal communities whose economies rely principally on their fishing fleet.

Indigenous Peoples

A Navajo man on horseback in Monument valley, Arizona.

Some Inuit people on a traditional *qamutik* (dog sled) in Cape Dorset, Nunavut, Canada.

Indigenous people, aboriginal people, or native people, are groups protected in international or national legislation as having a set of specific rights based on their linguistic and historical ties to a particular territory, their cultural and historical distinctiveness from other populations. The legislation is based on the conclusion that certain indigenous people are vulnerable to exploitation, marginalization, oppression, forced assimilation, and genocide by nation states formed from colonizing populations or by politically dominant, different ethnic groups.

A special set of political rights in accordance with international law have been set forth by international organizations such as the United Nations, the International Labour Organization and the World Bank. The United Nations has issued a Declaration on the Rights of Indigenous Peoples to guide member-state national policies to collective rights of indigenous people—such as culture, identity, language, and access to employment, health, education, and natural resources. Estimates put the total population of indigenous peoples from 220 million to 350 million.

A defining characteristic for an indigenous group is that it has preserved traditional ways of living, such as present or historical reliance upon subsistence-based production (based on pastoral, horticultural and/or hunting and gathering techniques), and a predominantly non-urbanized society. Not all indigenous groups share these characteristics. Indigenous societies may be either settled in a given locale/region or exhibit a nomadic lifestyle across a large territory, but are generally historically associated with a specific territory on which they depend. Indigenous societies are found in every inhabited climate zone and continent of the world.

Indigenous peoples are increasingly faced with threats to their sovereignty, environment, and access to natural resources. Examples of this can be the deforestation of tropical rainforests where several of the native tribe's subsistence and their normal lifestyle are threatened. Assimilative colonial policies resulted in ongoing issues related to aboriginal child protection.

Etymology

Etymology meaning "native" or "born within". Any given people, ethnic group or community may be described as *indigenous* in reference to some particular region or location that they see as their traditional tribal land claim. Other terms used to refer to indigenous populations are aboriginal, native, original, or first (as in Canada's First Nations).

The use of the term *peoples* in association with the indigenous is derived from the 19th century anthropological and ethnographic disciplines that Merriam-Webster Dictionary defines as "a body of persons that are united by a common culture, tradition, or sense of kinship, which typically have common language, institutions, and beliefs, and often constitute a politically organized group".

During the late twentieth century, the term *Indigenous people* began to be used to describe a legal category in indigenous law created in international and national legislation; it refers to culturally distinct groups affected by colonization.

James Anaya, former Special Rapporteur on the Rights of Indigenous Peoples, has defined indigenous peoples as "living descendants of pre-invasion inhabitants of lands now dominated by others. They are culturally distinct groups that find themselves engulfed by other settler societies born of forces of empire and conquest".

They form at present non-dominant sectors of society and are determined to preserve, develop and transmit to future generations their ancestral territories, and their ethnic identity, as the basis of their continued existence as peoples, in accordance with their own cultural patterns, social institutions and legal system. The International Day of the World's Indigenous People falls on 9 August as this was the date of the first meeting in 1982 of the United Nations Working Group of Indigenous Populations of the Subcommission on Prevention of Discrimination and Protection of Minorities of the Commission on Human Rights.

National Definitions

Ainu man of Hokkaidō, Japan in traditional dress

Throughout history, different states designate the groups within their boundaries that are recognized as indigenous peoples according to international or national legislation by different terms. Indigenous people also include people indigenous based on their descent from populations that inhabited the country when non-indigenous religions and cultures arrived—or at the establishment of present state boundaries—who retain some or all of their own social, economic, cultural and political institutions, but who may have been displaced from their traditional domains or who may have resettled outside their ancestral domains.

The status of the indigenous groups in the subjugated relationship can be characterized in most instances as an effectively marginalized, isolated or minimally participative one, in comparison to majority groups or the nation-state as a whole. Their ability to influence and participate in the external policies that may exercise jurisdiction over their traditional lands and practices is very frequently limited. This situation can persist even in the case where the indigenous population outnumbers that of the other inhabitants of the region or state; the defining notion here is one of separation from decision and regulatory processes that have some, at least titular, influence over aspects of their community and land rights.

In a ground-breaking 1997 decision involving the Ainu people of Japan, the Japanese courts recognised their claim in law, stating that "If one minority group lived in an area prior to being ruled over by a majority group and preserved its distinct ethnic culture even after being ruled over by the majority group, while another came to live in an area ruled over by a majority after consenting to the majority rule, it must be recognised that it is only natural that the distinct ethnic culture of the former group requires greater consideration."

The presence of external laws, claims and cultural mores either potentially or actually act to variously constrain the practices and observances of an indigenous society. These constraints can be observed even when the indigenous society is regulated largely by its own tradition and custom.

They may be purposefully imposed, or arise as unintended consequence of trans-cultural interaction. They may have a measurable effect, even where countered by other external influences and actions deemed beneficial or that promote indigenous rights and interests.

United Nations

In 1972 the United Nations Working Group on Indigenous Populations (WGIP) accepted as a preliminary definition a formulation put forward by Mr. José R. Martínez-Cobo, Special Rapporteur on Discrimination against Indigenous Populations. This definition has some limitations, because the definition applies mainly to pre-colonial populations, and would likely exclude other isolated or marginal societies.

Indigenous communities, peoples, and nations are those that, having a historical continuity with pre-invasion and pre-colonial societies that developed on their territories, consider themselves distinct from other sectors of the societies now prevailing in those territories, or parts of them. They form at present non-dominant sectors of society and are determined to preserve, develop, and transmit to future generations their ancestral territories, and their ethnic identity, as the basis of their continued existence as peoples, in accordance with their own cultural patterns, social institutions and legal systems.

The primary impetus in considering indigenous identity comes from the post-colonial movements and considering the historical impacts on populations by the European imperialism. The first paragraph of the Introduction of a report published in 2009 by the Secretariat of the Permanent Forum on Indigenous Issues published a report, states

For centuries, since the time of their colonization, conquest or occupation, indigenous peoples have documented histories of resistance, interface or cooperation with states, thus demonstrating their conviction and determination to survive with their distinct sovereign identities. Indeed, indigenous peoples were often recognized as sovereign peoples by states, as witnessed by the hundreds of treaties concluded between indigenous peoples and the governments of the United States, Canada, New Zealand and others.

History

Classical Antiquity

Greek sources of the Classical period acknowledge the prior existence of indigenous people(s), whom they referred to as "Pelasgians". These peoples inhabited lands surrounding the Aegean Sea before the subsequent migrations of the Hellenic ancestors claimed by these authors. The disposition and precise identity of this former group is elusive, and sources such as Homer, Hesiod and Herodotus give varying, partially mythological accounts. However, it is clear that cultures existed whose indigenous characteristics were distinguished by the subsequent Hellenic cultures (and distinct from non-Greek speaking "foreigners", termed "barbarians" by the historical Greeks).

Greco-Roman society flourished between 250 BC and 480 AD and commanded successive waves of conquests that gripped more than half of the globe. But because already existent populations within other parts of Europe at the time of classical antiquity had more in common culturally speaking with the Greco-Roman world, the intricacies involved in expansion across the European frontier were not so contentious relative to indigenous issues.

Alonso Fernández de Lugo presenting the captured Guanche kings of Tenerife to Ferdinand and Isabella

But when it came to expansion in other parts of the world, namely Asia, Africa, and the Middle East, then totally new cultural dynamics had entered into the equation, so to speak, and one sees here of what was to take the Americas, South East Asia, and the Pacific by storm a few hundred years later. The idea that peoples who possessed cultural customs and racial appearances strikingly different from those of the colonizing power is no new idea borne out of the Medieval period or the Enlightenment.

European Expansion and Colonialism

The rapid and extensive spread of the various European powers from the early 15th century onwards had a profound impact upon many of the indigenous cultures with whom they came into contact. The exploratory and colonial ventures in the Americas, Africa, Asia and the Pacific often resulted in territorial and cultural conflict, and the intentional or unintentional displacement and devastation of the indigenous populations.

The Canary Islands had an indigenous population called the Guanches whose origin is still the subject of discussion among historians and linguists.

Population and Distribution

Members of an uncontacted tribe encountered in the Brazilian state of Acre in 2009.

A Kawanua tribesman in a parade. Native Indonesians make up about 95% of 200 million Indonesian population.

Indigenous societies range from those who have been significantly exposed to the colonizing or expansionary activities of other societies (such as the Maya peoples of Mexico and Central America) through to those who as yet remain in comparative isolation from any external influence (such as the Sentinelese and Jarawa of the Andaman Islands).

Precise estimates for the total population of the world's Indigenous peoples are very difficult to compile, given the difficulties in identification and the variances and inadequacies of available census data. The United Nations estimates that there are over 370 million indigenous people living in over 70 countries worldwide. This would equate to just fewer than 6% of the total world population. This includes at least 5000 distinct peoples in over 72 countries.

Contemporary distinct indigenous groups survive in populations ranging from only a few dozen to hundreds of thousands and more. Many indigenous populations have undergone a dramatic decline and even extinction, and remain threatened in many parts of the world. Some have also been assimilated by other populations or have undergone many other changes. In other cases, indigenous populations are undergoing a recovery or expansion in numbers.

Certain indigenous societies survive even though they may no longer inhabit their "traditional" lands, owing to migration, relocation, forced resettlement or having been supplanted by other cultural groups. In many other respects, the transformation of culture of indigenous groups is ongoing, and includes permanent loss of language, loss of lands, encroachment on traditional territories, and disruption in traditional lifeways due to contamination and pollution of waters and lands.

Indigenous Peoples by Region

Indigenous populations are distributed in regions throughout the globe. The numbers, condition and experience of indigenous groups may vary widely within a given region. A comprehensive survey is further complicated by sometimes contentious membership and identification.

Africa

Tuareg nomads in southern Algeria

Starting fire by hand. San people in Botswana.

In the post-colonial period, the concept of specific indigenous peoples within the African continent has gained wider acceptance, although not without controversy. The highly diverse and numerous ethnic groups that comprise most modern, independent African states contain within them various peoples whose situation, cultures and pastoralist or hunter-gatherer lifestyles are generally marginalized and set apart from the dominant political and economic structures of the nation. Since the late 20th century these peoples have increasingly sought recognition of their rights as distinct indigenous peoples, in both national and international contexts.

Though the vast majority of African peoples are indigenous in the sense that they originate from that continent and middle and south east Asia—in practice, identity as an *indigenous people* per the modern definition is more restrictive, and certainly not every African ethnic group claims identification under these terms. Groups and communities who do claim this recognition are those who, by a variety of historical and environmental circumstances, have been placed outside of the dominant state systems, and whose traditional practices and land claims often come into conflict with the objectives and policies implemented by governments, companies and surrounding dominant societies.

Given the extensive and complicated history of human migration within Africa, being the "first peoples in a land" is not a necessary precondition for acceptance as an indigenous people. Rather, indigenous identity relates more to a set of characteristics and practices than priority of arrival. For example, several populations of nomadic peoples such as the Tuareg of the Sahara and Sahel regions now inhabit areas where they arrived comparatively recently; their claim to indigenous status (endorsed by the African Commission on Human and Peoples' Rights) is based on their marginalization as nomadic peoples in states and territories dominated by sedentary agricultural peoples.

Americas

Shaman from the shuara culture in Ecuador Amazonian forest

Quechua woman and child in the Sacred Valley, Andes, Peru

A Maya family in the hamlet of Patzutzun, Guatemala, 1993

Indigenous peoples of the American continent are broadly recognized as being those groups and their descendants who inhabited the region before the arrival of European colonizers and settlers (i.e., Pre-Columbian). Indigenous peoples who maintain, or seek to maintain, traditional ways of life are found from the high Arctic north to the southern extremities of Tierra del Fuego.

The impact of European colonization of the Americas on the indigenous communities has been in general quite severe, with many authorities estimating ranges of significant population decline primarily due to disease but also violence. The extent of this impact is the subject of much continuing debate. Several peoples shortly thereafter became extinct, or very nearly so.

All nations in North and South America have populations of indigenous peoples within their borders. In some countries (particularly Latin American), indigenous peoples form a sizable component of the overall national population—in Bolivia they account for an estimated 56%–70% of the total nation, and at least half of the population in Guatemala and the Andean and Amazonian nations of Peru. In English, indigenous peoples are collectively referred to by different names that vary by region and include such ethnonyms as Native Americans, Amerindians, and American Indians. In Spanish or Portuguese speaking countries one finds the use of terms such as *pueblos indígenas*, *amerindios*, *povos nativos*, *povos indígenas*, and in Peru, *Comunidades Nativas* (Native Communities), particularly among Amazonian societies like the Urarina and Matsés. In Chile there are indigenous tribes like the Mapuches in the Center-South and the Aymaras in the North, also the Rapa Nui indigenous to Easter Island are a Polynesian tribe.

In Brazil, the term *índio* is used by most of the population, the media, the indigenous peoples themselves and even the government (National Indio Foundation), although its Hispanic equivalent *indio* is widely not considered politically correct and falling into disuse.

Raoni Metuktire, Kaye, Kadjor and Panara, leaders of the indigenous Kayapo tribe, Mato Grosso, Brazil

Aboriginal peoples in Canada comprise the First Nations, Inuit and Métis. The descriptors "Indian" and "Eskimo" are falling into disuse in Canada. There are currently over 600 recognized First Nations governments or bands encompassing 1,272,790 peoples spread across Canada with distinctive Aboriginal cultures, languages, art, and music. National Aboriginal Day recognises the cultures and contributions of Aboriginals to the history of Canada

The Inuit have achieved a degree of administrative autonomy with the creation in 1999 of the territories of Nunavik (in Northern Québec), Nunatsiavut (in Northern Labrador) and Nunavut, which was until 1999 a part of the Northwest Territories. The self-ruling Danish territory of Greenland is also home to a majority population of indigenous Inuit (about 85%).

In the United States, the combined populations of Native Americans, Inuit and other indigenous designations totalled 2,786,652 (constituting about 1.5% of 2003 US census figures). Some 563 scheduled tribes are recognized at the federal level, and a number of others recognized at the state level.

In Mexico, approximately 6,011,202 (constituting about 6.7% of 2005 Mexican census figures) identify as *Indígenas* (Spanish for natives or indigenous peoples). In the southern states of Chiapas, Yucatán and Oaxaca they constitute 26.1%, 33.5% and 35.3%, respectively, of the population. In these states several conflicts and episodes of civil war have been conducted, in which the situation and participation of indigenous societies were notable factors.

The Amerindians make up 0.4% of all Brazilian population, or about 700,000 people. Indigenous peoples are found in the entire territory of Brazil, although the majority of them live in Indian reservations in the North and Center-Western part of the country. On 18 January 2007, FUNAI reported that it had confirmed the presence of 67 different uncontacted tribes in Brazil, up from 40 in 2005. With this addition Brazil has now overtaken the island of New Guinea as the country having the largest number of uncontacted tribes.

Asia

The Circassians are one of the oldest nations in the European North Caucasus.

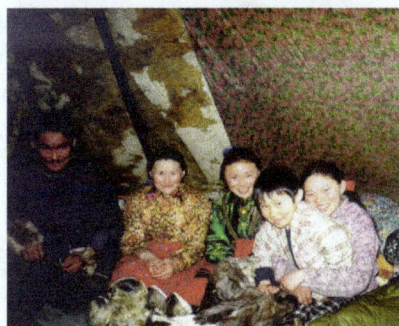

A Nenets family in their tent, Yamal peninsula, Russia.

Dayak man from Indonesia, Southeast Asia.

The vast regions of Asia contain the majority of the world's present-day Indigenous populations, about 70% according to IWGIA figures.

The indigenous peoples of the Chittagong Hill Tracts are the Buddhist Chakma people (Jumma people).

The most substantial populations are in India, which constitutionally recognizes a range of "Scheduled Tribes" within its borders. These various peoples (collectively referred to as Adivasis, or tribal peoples) number about 68 million (1991 census figures, approximately 8% of the total national population).

There are also indigenous people residing in the hills of Northern, North-eastern and Southern India like the Ladakhi, Kinnaurs, Lepcha, Bhutia (of Sikkim), Naga (of Nagaland), Bodo, Munda people of Chota Nagpur Plateau, Mizo (of Mizoram), Kodava (of Kodagu), Toda, Kurumba, Kota (of the Nilgiris), Irulas and others.

The Russians invaded Siberia and conquered the indigenous natives in the 17th-18th centuries.

Nivkh people are an ethnic group indigenous to Sakhalin, having a few speakers of the Nivkh language, but their fisher culture has been endangered due to the development of oil field of Sakhalin from 1990s.

Ainu people are an ethnic group indigenous to Hokkaidō, the Kuril Islands, and much of Sakhalin. As Japanese settlement expanded, the Ainu were pushed northward and fought against the Japanese in Shakushain's Revolt and Menashi-Kunashir Rebellion, until by the Meiji period they were confined by the government to a small area in Hokkaidō, in a manner similar to the placing of Native Americans on reservations.

The Ryukyuan people are indigenous to the Ryukyu Islands.

The Dzungar Oirats are the natives of Dzungaria in Northern Xinjiang.

The Pamiris are the native people of Tashkurgan in Xinjiang.

The Yazidis are indigenous to the Sinjar mountain range in northern Iraq.

The languages of Taiwanese aborigines have significance in historical linguistics, since in all likelihood Taiwan was the place of origin of the entire Austronesian language family, which spread across Oceania.

There are also indigenous people in Southeast Asia. There are indigenous peoples of the Philippines, which Spain and the United States colonized.

The Javanese, Sundanese, Bantenese, Betawi, Tengger, Osing, Badui, Madurese, Malays, Batak, Minangkabau, Acehnese, Lampung, Kubu, Dayak, Banjar, Makassarese, Buginese, Mandar, Minahasa, Buton, Gorontalo, Toraja, Bajau, Balinese, Sasak, Nuaulu, Manusela, Wemale, Dani, Bauzi, Asmat are indigenous peoples in Indonesia. There are over 300 ethnic groups in Indonesia. 200 million of those are of Native Indonesian ancestry.

The Cham are the indigenous people of the former state of Champa which was conquered by Vietnam in the Cham–Vietnamese wars during Nam tiến. The Cham in Vietnam are only recognized as a minority, and not as an indigenous people by the Vietnamese government despite being indigenous to the region.

The Degar (Montagnards) are the natives of the Central Highlands (Vietnam) and were conquered by the Vietnamese in the Nam tiến.

The Khmer Krom are the native people of the Mekong Delta and Saigon which were acquired by Vietnam from Cambodian King Chey Chettha II in exchange for a Vietnamese princess.

The indigenous people of Cordillera Administrative Region in the Philippines are the Igorot people.

The indigenous peoples of Mindanao are the Lumad peoples and the Moro (Tausug, Maguindanao Maranao and others) who also live in the Sulu archipelago.

The indigenous people of Northern Iraq are the Assyrians. They descended from the ancient Neo-Assyrian Empire and Akkadians, and lived in what was Assyria, their original homeland.

The Jats are indigenous people of [ancient] India, and can be tracked down to 4th century BC.

Europe

Sven-Roald Nystø, Aili Keskitalo and Ole Henrik Magga, the three first presidents of the Norwegian Sami Parliament

In Europe, present-day indigenous populations as recognized by the UN are relatively few, mainly confined to northern and far-eastern reaches of this Eurasian peninsula. Nevertheless, the ethnic groups traditionally inhabiting most, if not all, European countries are considered to be indigenous to Europe. This includes the majority populations. It can lead to some confusion that the term "indigenous" does not imply "non-white" or "minority" in Europe, as it would in the Americas and Australia.

Notable minority indigenous populations in Europe include the Basque people of northern Spain and southern France, the Sami people of northern Scandinavia, the Nenets and other Samoyedic peoples of the northern Russian Federation, and the Komi peoples of the western Urals, beside the Circassians in the North Caucasus.

Oceania

Huli man from the Southern Highlands, Papua New Guinea. New Guinea has more than 1,000 indigenous languages.

In Australia the indigenous populations are the Aboriginal Australians, within which are many different nations and tribes, and the Torres Strait Islanders. These groups are often spoken of as Indigenous Australians.

Many of the present-day Pacific Island nations in the Oceania region were originally populated by Polynesian, Melanesian and Micronesian peoples over the course of thousands of years. European colonial expansion in the Pacific brought many of these under non-indigenous administration. During the 20th century several of these former colonies gained independence and nation-states were formed under local control. However, various peoples have put forward claims for Indigenous recognition where their islands are still under external administration; examples include the Chamorros of Guam and the Northern Marianas, and the Marshallese of the Marshall Islands.

The remains of at least 25 miniature humans, who lived between 1,000 and 3,000 years ago, were recently found on the islands of Palau in Micronesia.

In most parts of Oceania, indigenous peoples outnumber the descendants of colonists. Exceptions include New Zealand and Hawaii. According to the 2013 census, New Zealand Māori make up 14.9% of the population of New Zealand, with less than half (46.5%) of all Māori residents identifying solely as Māori. The Māori are indigenous to Polynesia and settled New Zealand relatively recently, the migrations were thought to have occurred in the 13th century CE. In New Zealand pre-contact Māori tribes were not a single people, thus the more recent grouping into tribal (iwi) arrangements has become a more formal arrangement in more recent times. Many Māori tribal leaders signed a treaty with the British, the Treaty of Waitangi, which formed the modern geo-political entity that is New Zealand.

The independent state of Papua New Guinea (PNG) has a majority population of indigenous societies, with more than 700 different tribal groups recognized out of a total population of just over 5 million. The PNG Constitution and other Acts identify traditional or custom-based practices and land tenure, and explicitly set out to promote the viability of these traditional societies within the modern state. However, conflicts and disputes concerning land use and resource rights continue between indigenous groups, the government, and corporate entities.

Indigenous Rights and Other Issues

The New Zealand delegation endorses the United Nations Declaration on the Rights of Indigenous Peoples in April 2010.

Indigenous peoples confront a diverse range of concerns associated with their status and interaction with other cultural groups, as well as changes in their inhabited environment. Some challenges are specific to particular groups; however, other challenges are commonly experienced. These issues include cultural and linguistic preservation, land rights, ownership and exploitation of natural resources, political determination and autonomy, environmental degradation and incursion, poverty, health, and discrimination.

The interaction between indigenous and non-indigenous societies throughout history has been complex, ranging from outright conflict and subjugation to some degree of mutual benefit and cultural transfer. A particular aspect of anthropological study involves investigation into the ramifications of what is termed *first contact*, the study of what occurs when two cultures first encounter one another. The situation can be further confused when there is a complicated or contested history of migration and population of a given region, which can give rise to disputes about primacy and ownership of the land and resources.

Wherever indigenous cultural identity is asserted, common societal issues and concerns arise from the indigenous status. These concerns are often not unique to indigenous groups. Despite the diversity of Indigenous peoples, it may be noted that they share common problems and issues in dealing with the prevailing, or invading, society. They are generally concerned that the cultures of Indigenous peoples are being lost and that indigenous peoples suffer both discrimination and pressure to assimilate into their surrounding societies. This is borne out by the fact that the lands and cultures of nearly all of the peoples listed at the end of this article are under threat. Notable exceptions are the Sakha and Komi peoples (two of the northern indigenous peoples of Russia), who now control their own autonomous republics within the Russian state, and the Canadian Inuit, who form a majority of the territory of Nunavut (created in 1999). In Australia, a landmark case, Mabo v Queensland (No 2), saw the High Court of Australia reject the idea of terra nullius. This rejection ended up recognizing that there was a pre-existing system of law practiced by the Meriam people.

It is also sometimes argued that it is important for the human species as a whole to preserve a wide range of cultural diversity as possible, and that the protection of indigenous cultures is vital to this enterprise.

Human Rights Violation

The Bangladesh Government has stated that there are "no Indigenous Peoples in Bangladesh". This has angered the Indigenous Peoples of Chittagong Hill Tracts, Bangladesh, collectively known as the Jumma. Experts have protested against this move of the Bangladesh Government and have questioned the Government's definition of the term "Indigenous Peoples". This move by the Bangladesh Government is seen by the Indigenous Peoples of Bangladesh as another step by the Government to further erode their already limited rights.

Both Hindu and Chams have experienced religious and ethnic persecution and restrictions on their faith under the current Vietnamese government, with the Vietnamese state confisticating Cham property and forbidding Cham from observing their religious beliefs. Hindu temples were turned into tourist sites against the wishes of the Cham Hindus. In 2010 and 2013 several incidents occurred in Thành Tín and Phước Nhơn villages where Cham were murdered by Vietnamese. In 2012, Vietnamese police in Chau Giang village stormed into a Cham Mosque, stole the electric generator, and also raped Cham girls. Cham in the Mekong Delta have also been economically marginalised, with ethnic Vietnamese settling on land previously owned by Cham people with state support.

The French, the Communist North Vietnamese, and the anti-Communist South Vietnamese all exploited and persecuted the Montagnards. North Vietnamese Communists forcibly recruited "comfort girls" from the indigenous Montagnard peoples of the Central Highlands and murdered those who didn't comply, inspired by Japan's use of comfort women. The Vietnamese viewed and dealt with the indigenous Montagnards in the CIDG from the Central Highlands as "savages" and this caused a Montagnard uprising against the Vietnamese. The Vietnamese were originally centered around the Red River Delta but engaged in conquest and seized new lands such as Champa, the

Mekong Delta (from Cambodia) and the Central Highlands during Nam Tien, while the Vietnamese received strong Chinese influence in their culture and civilization and were Sinicized, and the Cambodians and Laotians were Indianized, the Montagnards in the Central Highlands maintained their own native culture without adopting external culture and were the true indigenous natives of the region, and to hinder encroachment on the Central Highlands by Vietnamese nationalists, the term *Pays Montagnard du Sud-Indochinois* PMSI emerged for the Central Highlands along with the natives being addressed by the name Montagnard. The tremendous scale of Vietnamese Kinh colonists flooding into the Central Highlands has significantly altered the demographics of the region. The anti-ethnic minority discriminatory policies by the Vietnamese, environmental degradation, deprivation of lands from the natives, and settlement of native lands by a massive amount of Vietnamese settlers led to massive protests and demonstrations by the Central Highland's indigenous native ethnic minorities against the Vietnamese in January–February 2001 and this event gave a tremendous blow to the claim often published by the Vietnamese government that in Vietnam *There has been no ethnic confrontation, no religious war, no ethnic conflict. And no elimination of one culture by another.*

Health Issues

In December 1993, the United Nations General Assembly proclaimed the International Decade of the World's Indigenous People, and requested UN specialized agencies to consider with governments and indigenous people how they can contribute to the success of the Decade of Indigenous People, commencing in December 1994. As a consequence, the World Health Organization, at its Forty-seventh World Health Assembly established a core advisory group of indigenous representatives with special knowledge of the health needs and resources of their communities, thus beginning a long-term commitment to the issue of the health of indigenous peoples.

The WHO notes that "Statistical data on the health status of indigenous peoples is scarce. This is especially notable for indigenous peoples in Africa, Asia and eastern Europe", but snapshots from various countries, where such statistics are available, show that indigenous people are in worse health than the general population, in advanced and developing countries alike: higher incidence of diabetes in some regions of Australia; higher prevalence of poor sanitation and lack of safe water among Twa households in Rwanda; a greater prevalence of childbirths without prenatal care among ethnic minorities in Vietnam; suicide rates among Inuit youth in Canada are eleven times higher than the national average; infant mortality rates are higher for indigenous peoples everywhere.

Indigenous Worldviews and The Global Community

International relations (IR) is inherently one born from the Western European modes of thought, and from a view of its early structures and theoretical foundations, including the work of European scholars Hobbes, Locke, and Rousseau, it fails to incorporate Indigenous understandings, which consequently further perpetuate colonization rather than prevent it. International organizations are founded under theoretical IR approaches, and assume the nature of humans, but contradict with the assumption of the nature of humans being inherently competitive and therefore draw conclusions that the Hobbesian "Social Contract" should not exist in International bodies today.

The contradiction seen when comparing the western European tradition of defining human nature to western European nature itself has impacted the ways at which colonial states, like Canada, interact with Indigenous peoples whom became subject to western European models of thought in respect to the definition of human nature, and therefore pure theoretics regarding IR in respect to colonial-Indigenous relations falls short. The shortcoming is mainly due to the vast assumptions made regarding the nature of humans, which fail to consider Indigenous forms of "international relations". The treaty process during the years of pioneer settlement for example served as legal weapons by the western European world to lay claim through its own laws and understands to the vast territory that Indigenous people have been in relationship with since time immemorial. "As exemplified in jurisprudential, statute, and constitutional law, Canada imagines that indigenous peoples have already been incorporated into the state. That is, the Canadian state assumes that indigenous peoples already come under Canadian political jurisdiction". The problem then, in summary, is that IR's failed ability to relate in an intimate way with the worldviews and teaching that govern Indigenous people is not really International Relations but rather just another naturalized colonial tactic perhaps even un-beware to its beholder. A failure to incorporate Indigenous understandings in the overarching western-European founded study of IR isn't the problem of Indigenous peoples not being able to adapt or engage with its colonizer as a recognized Westphalia state but that the failed assumption that the western European foundational models of human nature are correct. IR's foundational theories then serve better in understanding western European states of nature.

A key difference in the models of knowledge from an Indigenous worldview and that of the western European founded model is the ways at which both groups behave as collectives in response to one another; cooperation verse competition respectively. In the nature of cooperation women, children, elders, men, all members of society have a place in building the way forward for generations. In the nature of competition only the strongest or those with the means to security have a place in society and those outside that privilege become wards of the protectorate. Ways of contemporary decolonization seek to establish the legitimacy and un-naturalize the assumptions of the nature of humans and institute rather many forms or understanding ourselves, Indigenous and non- Indigenous.

Non-indigenous Viewpoints

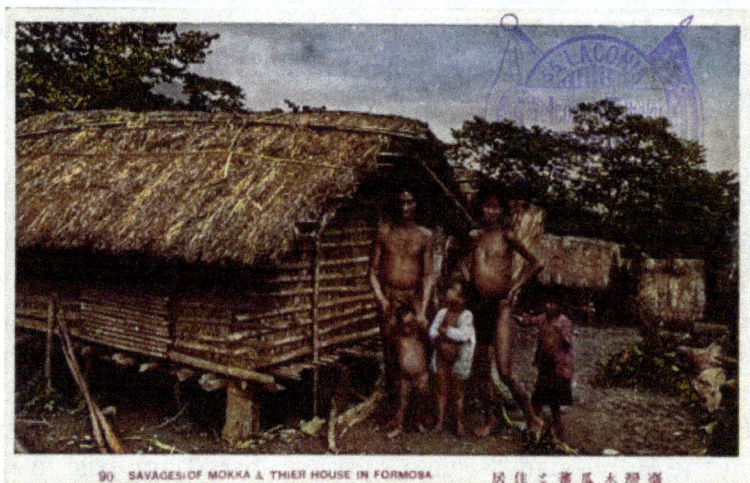

"Savages of Mokka and Their House in Formosa", pre-1945, Taiwan under Japanese rule.

Indigenous peoples have been denoted *primitives*, *savages*, or *uncivilized*. These terms were common during the heights of European colonial expansion, but still continue in modern times.

During the 17th century, indigenous peoples were commonly labeled "uncivilized". Some philosophers such as Thomas Hobbes considered indigenous people to be merely 'savages', while others are purported to have considered them to be "noble savages". Those who were close to the Hobbesian view tended to believe themselves to have a duty to "civilize" and "modernize" the indigenous. Although anthropologists, especially from Europe, used to apply these terms to all tribal cultures, it has fallen into disfavor as demeaning and is, according to many anthropologists, not only inaccurate, but dangerous.

Survival International runs a campaign to stamp out media portrayal of indigenous peoples as 'primitive' or 'savages'. Friends of Peoples Close to Nature considers not only that indigenous culture should be respected as not being inferior, but also sees their way of life as a lesson of sustainability and a part of the struggle within the "corrupted" western world, from which the threat stems.

After World War I, however, many Europeans came to doubt the morality of the means used to "civilize" peoples. At the same time, the anti-colonial movement, and advocates of indigenous peoples, argued that words such as "civilized" and "savage" were products and tools of colonialism, and argued that colonialism itself was savagely destructive. In the mid 20th century, European attitudes began to shift to the view that indigenous and tribal peoples should have the right to decide for themselves what should happen to their ancient cultures and ancestral lands.

Treaty

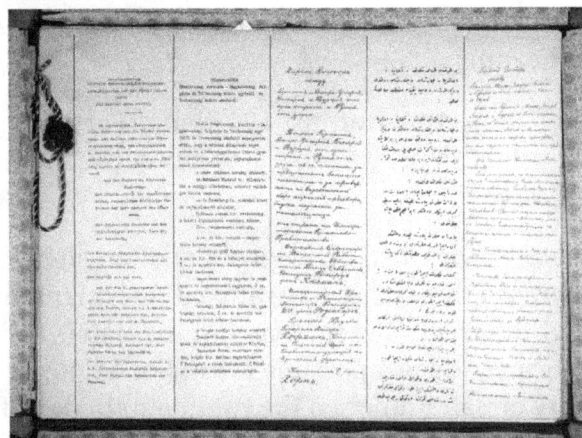

The first two pages of the Treaty of Brest-Litovsk, in (left to right) German, Hungarian, Bulgarian, Ottoman Turkish and Russian

A treaty is an agreement under international law entered into by actors in international law, namely sovereign states and international organizations. A treaty may also be known as an (international) agreement, protocol, covenant, convention, pact, or exchange of letters, among other terms. Regardless of terminology, all of these forms of agreements are, under international law, equally considered treaties and the rules are the same.

Treaties can be loosely compared to contracts: both are means of willing parties assuming obligations among themselves, and a party to either that fails to live up to their obligations can be held liable under international law.

Modern Usage

A treaty is an official, express written agreement that states use to legally bind themselves. A treaty is the official document which expresses that agreement in words; and it is also the objective outcome of a ceremonial occasion which acknowledges the parties and their defined relationships.

Modern Form

Since the late 19th century, most treaties have followed a fairly consistent format. A treaty typically begins with a preamble describing the contracting parties and their joint objectives in executing the treaty, as well as summarizing any underlying events (such as a war). Modern preambles are sometimes structured as a single very long sentence formatted into multiple paragraphs for readability, in which each of the paragraphs begins with a verb (desiring, recognizing, having, and so on).

The contracting parties' full names or sovereign titles are often included in the preamble, along with the full names and titles of their representatives, and a boilerplate clause about how their representatives have communicated (or exchanged) their full powers (i.e., the official documents appointing them to act on behalf of their respective states) and found them in good or proper form.

The end of the preamble and the start of the actual agreement is often signaled by the words "have agreed as follows."

After the preamble comes numbered articles, which contain the substance of the parties' actual agreement. Each article heading usually encompasses a paragraph. A long treaty may further group articles under chapter headings.

Modern treaties, regardless of subject matter, usually contain articles governing where the final authentic copies of the treaty will be deposited and how any subsequent disputes as to their interpretation will be peacefully resolved.

The end of a treaty, the eschatocol (or closing protocol), is often signaled by a clause like "in witness whereof" or "in faith whereof," the parties have affixed their signatures, followed by the words "DONE at," then the site(s) of the treaty's execution and the date(s) of its execution. The date is typically written in its most formal, longest possible form. For example, the Charter of the United Nations was "DONE at the city of San Francisco the twenty-sixth day of June, one thousand nine hundred and forty-five." If the treaty is executed in multiple copies in different languages, that fact is always noted, and is followed by a stipulation that the versions in different languages are equally authentic.

The signatures of the parties' representatives follow at the very end. When the text of a treaty is later reprinted, such as in a collection of treaties currently in effect, an editor will often append the dates on which the respective parties ratified the treaty and on which it came into effect for each party.

Bilateral and Multilateral Treaties

Bilateral treaties are concluded between two states or entities. It is possible, however, for a bilateral treaty to have more than two parties; consider for instance the bilateral treaties between Switzerland and the European Union (EU) following the Swiss rejection of the European Economic Area agreement. Each of these treaties has seventeen parties. These however are still bilateral, not multilateral, treaties. The parties are divided into two groups, the Swiss ("on the one part") and the EU and its member states ("on the other part"). The treaty establishes rights and obligations between the Swiss and the EU and the member states severally—it does not establish any rights and obligations amongst the EU and its member states.

A multilateral treaty is concluded among several countries. The agreement establishes rights and obligations between each party and every other party. Multilateral treaties are often regional. Treaties of "mutual guarantee" are international compacts, e.g., the Treaty of Locarno which guarantees each signatory against attack from another.

Adding and Amending Treaty Obligations

Reservations

Reservations are essentially caveats to a state's acceptance of a treaty. Reservations are unilateral statements purporting to exclude or to modify the legal obligation and its effects on the reserving state. These must be included at the time of signing or ratification, i.e. "a party cannot add a reservation after it has already joined a treaty".

Originally, international law was unaccepting of treaty reservations, rejecting them unless all parties to the treaty accepted the same reservations. However, in the interest of encouraging the largest number of states to join treaties, a more permissive rule regarding reservations has emerged. While some treaties still expressly forbid any reservations, they are now generally permitted to the extent that they are not inconsistent with the goals and purposes of the treaty.

When a state limits its treaty obligations through reservations, other states party to that treaty have the option to accept those reservations, object to them, or object and oppose them. If the state accepts them (or fails to act at all), both the reserving state and the accepting state are relieved of the reserved legal obligation as concerns their legal obligations to each other (accepting the reservation does not change the accepting state's legal obligations as concerns other parties to the treaty). If the state opposes, the parts of the treaty affected by the reservation drop out completely and no longer create any legal obligations on the reserving and accepting state, again only as concerns each other. Finally, if the state objects and opposes, there are no legal obligations under that treaty between those two state parties whatsoever. The objecting and opposing state essentially refuses to acknowledge the reserving state is a party to the treaty at all.

Amendments

There are three ways an existing treaty can be amended. First, formal amendment requires State parties to the treaty to go through the ratification process all over again. The re-negotiation of treaty provisions can be long and protracted, and often some parties to the original treaty will not become parties to the amended treaty. When determining the legal obligations of states, one party to the original treaty and one a party to the amended treaty, the states will only be bound by the terms they both agreed upon. Treaties can also be amended informally by the treaty executive council when the changes are only procedural, technical change in customary international law can also amend a treaty, where state behavior evinces a new interpretation of the legal obligations under the treaty. Minor corrections to a treaty may be adopted by a procès-verbal; but a procès-verbal is generally reserved for changes to rectify obvious errors in the text adopted, i.e. where the text adopted does not correctly reflect the intention of the parties adopting it.

Protocols

In international law and international relations, a protocol is generally a treaty or international agreement that supplements a previous treaty or international agreement. A protocol can amend the previous treaty, or add additional provisions. Parties to the earlier agreement are not required to adopt the protocol. Sometimes this is made clearer by calling it an "optional protocol", especially where many parties to the first agreement do not support the protocol.

Some examples: the United Nations Framework Convention on Climate Change (UNFCCC) established a framework for the development of binding greenhouse gas emission limits, while the Kyoto Protocol contained the specific provisions and regulations later agreed upon.

Execution and Implementation

Treaties may be seen as 'self-executing', in that merely becoming a party puts the treaty and all of its obligations in action. Other treaties may be non-self-executing and require 'implementing legislation'—a change in the domestic law of a state party that will direct or enable it to fulfill treaty obligations. An example of a treaty requiring such legislation would be one mandating local prosecution by a party for particular crimes.

The division between the two is often not clear and is often politicized in disagreements within a government over a treaty, since a non-self-executing treaty cannot be acted on without the proper change in domestic law. If a treaty requires implementing legislation, a state may be in default of its obligations by the failure of its legislature to pass the necessary domestic laws.

Interpretation

The language of treaties, like that of any law or contract, must be interpreted when the wording does not seem clear or it is not immediately apparent how it should be applied in a perhaps unforeseen circumstance. The Vienna Convention states that treaties are to be interpreted "in good faith" according to the "ordinary meaning given to the terms of the treaty in their context and in the light of its object and purpose." International legal experts also often invoke the 'principle of maximum effectiveness,' which interprets treaty language as having the fullest force and effect possible to establish obligations between the parties.

No one party to a treaty can impose its particular interpretation of the treaty upon the other parties. Consent may be implied, however, if the other parties fail to explicitly disavow that initially unilateral interpretation, particularly if that state has acted upon its view of the treaty without complaint. Consent by all parties to the treaty to a particular interpretation has the legal effect of adding another clause to the treaty – this is commonly called an 'authentic interpretation'.

International tribunals and arbiters are often called upon to resolve substantial disputes over treaty interpretations. To establish the meaning in context, these judicial bodies may review the preparatory work from the negotiation and drafting of the treaty as well as the final, signed treaty itself.

Consequences of Terminology

One significant part of treaty making is that signing a treaty implies recognition that the other side is a sovereign state and that the agreement being considered is enforceable under international law. Hence, nations can be very careful about terming an agreement to be a treaty. For example, within the United States, agreements between states are compacts and agreements between states and the federal government or between agencies of the government are memoranda of understanding.

Another situation can occur when one party wishes to create an obligation under international law, but the other party does not. This factor has been at work with respect to discussions between North Korea and the United States over security guarantees and nuclear proliferation.

The terminology can also be confusing because a treaty may and usually is named something other than a treaty, such as a convention, protocol, or simply agreement. Conversely some legal documents such as the Treaty of Waitangi are internationally considered to be documents under domestic law.

Ending Treaty Obligations

Withdrawal

Treaties are not necessarily permanently binding upon the signatory parties. As obligations in international law are traditionally viewed as arising only from the consent of states, many treaties expressly allow a state to withdraw as long as it follows certain procedures of notification. For example, the Single Convention on Narcotic Drugs provides that the treaty will terminate if, as a result of denunciations, the number of parties falls below 40. Many treaties expressly forbid withdrawal. Article 56 of the Vienna Convention on the Law of Treaties provides that where a treaty is silent over whether or not it can be denounced there is a rebuttable presumption that it cannot be unilaterally denounced unless:

- it can be shown that the parties intended to admit the possibility, or

- a right of withdrawal can be inferred from the terms of the treaty.

The possibility of withdrawal depends on the terms of the treaty and its *travaux preparatoire*. It has, for example, been held that it is not possible to withdraw from the International Covenant on Civil and Political Rights. When North Korea declared its intention to do this the Secretary-General of the United Nations, acting as registrar, said that original signatories of the ICCPR had not overlooked the possibility of explicitly providing for withdrawal, but rather had deliberately intended not to provide for it. Consequently, withdrawal was not possible.

In practice, because of sovereignty, any state can purport to withdraw from any treaty at any time, and cease to abide by its terms. The question of whether this is lawfully might be regarded as really a question of how other states will react; for instance, another state might impose sanctions or go to war over a treaty violation.

If a state party's withdrawal is successful, its obligations under that treaty are considered terminated, and withdrawal by one party from a bilateral treaty of course terminates the treaty. When a state withdraws from a multi-lateral treaty, that treaty will still otherwise remain in force among the other parties, unless, of course, otherwise should or could be interpreted as agreed upon between the remaining states parties to the treaty.

Suspension and Termination

If a party has materially violated or breached its treaty obligations, the other parties may invoke this breach as grounds for temporarily suspending their obligations to that party under the treaty. A material breach may also be invoked as grounds for permanently terminating the treaty itself.

A treaty breach does not automatically suspend or terminate treaty relations, however. It depends on how the other parties regard the breach and how they resolve to respond to it. Sometimes treaties will provide for the seriousness of a breach to be determined by a tribunal or other independent arbiter. An advantage of such an arbiter is that it prevents a party from prematurely and perhaps wrongfully suspending or terminating its own obligations due to another's alleged material breach.

Treaties sometimes include provisions for self-termination, meaning that the treaty is automatically terminated if certain defined conditions are met. Some treaties are intended by the parties to be only temporarily binding and are set to expire on a given date. Other treaties may self-terminate if the treaty is meant to exist only under certain conditions.

A party may claim that a treaty should be terminated, even absent an express provision, if there has been a fundamental change in circumstances. Such a change is sufficient if unforeseen, if it undermined the "essential basis" of consent by a party, if it radically transforms the extent of obligations between the parties, and if the obligations are still to be performed. A party cannot base this claim on change brought about by its own breach of the treaty. This claim also cannot be used to invalidate treaties that established or redrew political boundaries.

Invalid Treaties

There are several reasons an otherwise valid and agreed upon treaty may be rejected as a binding international agreement, most of which involve problems created at the formation of the treaty. For example, the serial Japan-Korea treaties of 1905, 1907 and 1910 were protested; and they were confirmed as "already null and void" in the 1965 Treaty on Basic Relations between Japan and the Republic of Korea.

Ultra Vires Treaties

A party's consent to a treaty is invalid if it had been given by an agent or body without power to do so under that state's domestic law. States are reluctant to inquire into the internal affairs and processes of other states, and so a "manifest violation" is required such that it would be "objectively evident to any State dealing with the matter". A strong presumption exists internationally that a head of state has acted within his proper authority. It seems that no treaty has ever actually been invalidated on this provision.

Consent is also invalid if it is given by a representative who ignored restrictions he is subject to by his sovereign during the negotiations, if the other parties to the treaty were notified of those restrictions prior to his signing.

According to the preamble in The Law of Treaties, treaties are a source of international law. If an act or lack thereof is condemned under international law, the act will not assume international legality even if approved by internal law. This means that in case of a conflict with domestic law, international law will always prevail.

Misunderstanding, Fraud, Corruption, Coercion

Articles 46–53 of the Vienna Convention on the Law of Treaties set out the only ways that treaties can be invalidated—considered unenforceable and void under international law. A treaty will be invalidated due to either the circumstances by which a state party joined the treaty, or due to the content of the treaty itself. Invalidation is separate from withdrawal, suspension, or termination (addressed above), which all involve an alteration in the consent of the parties of a previously valid treaty rather than the invalidation of that consent in the first place.

A state's consent may be invalidated if there was an erroneous understanding of a fact or situation at the time of conclusion, which formed the "essential basis" of the state's consent. Consent will not be invalidated if the misunderstanding was due to the state's own conduct, or if the truth should have been evident.

Consent will also be invalidated if it was induced by the fraudulent conduct of another party, or by the direct or indirect "corruption" of its representative by another party to the treaty. Coercion of either a representative, or the state itself through the threat or use of force, if used to obtain the consent of that state to a treaty, will invalidate that consent.

Contrary to Peremptory Norms

A treaty is null and void if it is in violation of a peremptory norm. These norms, unlike other principles of customary law, are recognized as permitting no violations and so cannot be altered through treaty obligations. These are limited to such universally accepted prohibitions as those against the aggressive use of force, genocide and other crimes against humanity, piracy, hostilities directed at civilian population, racial discrimination and apartheid, slavery and torture, meaning that no state can legally assume an obligation to commit or permit such acts.

Role of The United Nations

The United Nations Charter states that treaties must be registered with the UN to be invoked before it or enforced in its judiciary organ, the International Court of Justice. This was done to prevent the proliferation of secret treaties that occurred in the 19th and 20th century. Section 103 of the Charter also states that its members' obligations under it outweigh any competing obligations under other treaties.

After their adoption, treaties as well as their amendments have to follow the official legal procedures of the United Nations, as applied by the Office of Legal Affairs, including signature, ratification and entry into force.

In function and effectiveness, the UN has been compared to the pre-Constitutional United States Federal government by some, giving a comparison between modern treaty law and the historical Articles of Confederation.

Relation Between National Law and Treaties by Country

Brazilian Law

The Brazilian federal constitution states that the power to enter into treaties is vested in the president and that such treaties must be approved by Congress (articles 84, clause VIII, and 49, clause I). In practice, this has been interpreted as meaning that the executive branch is free to negotiate and sign a treaty, but its ratification by the president is contingent upon the prior approval of Congress. Additionally, the Federal Supreme Court has ruled that, following ratification and entry into force, a treaty must be incorporated into domestic law by means of a presidential decree published in the federal register in order to be valid in Brazil and applicable by the Brazilian authorities.

The Federal Supreme Court has established that treaties are subject to constitutional review and enjoy the same hierarchical position as ordinary legislation (*leis ordinárias*, or "ordinary laws", in Portuguese). A more recent ruling by the Supreme Court in 2008 has altered that scheme somewhat, by stating that treaties containing human rights provisions enjoy a status above that of ordinary legislation, though they remain beneath the constitution itself. Additionally, as per the 45th amendment to the constitution, human rights treaties which are approved by Congress by means of a special procedure enjoy the same hierarchical position as a constitutional amendment. The hierarchical position of treaties in relation to domestic legislation is of relevance to the discussion on whether (and how) the latter can abrogate the former and vice versa.

The Brazilian federal constitution does not have a supremacy clause with the same effects as the one on the U.S. constitution, a fact that is of interest to the discussion on the relation between treaties and state legislation.

United States Law

In the United States, the term "treaty" has a different, more restricted legal sense than exists in international law. United States law distinguishes what it calls treaties from executive agreement, congressional-executive agreements, and sole executive agreements. All four classes are equally treaties under international law; they are distinct only from the perspective of internal American law. The distinctions are primarily concerning their method of approval. Whereas treaties require advice and consent by two-thirds of the Senators present, sole executive agreements may be executed by the President acting alone. Some treaties grant the President the authority to fill in the gaps with executive agreements, rather than additional treaties or protocols. And finally, congressional-executive agreements require majority approval by both the House and the Senate, either before or after the treaty is signed by the President.

Currently, international agreements are executed by executive agreement rather than treaties at a rate of 10:1. Despite the relative ease of executive agreements, the President still often chooses to pursue the formal treaty process over an executive agreement in order to gain congressional support on matters that require the Congress to pass implementing legislation or appropriate funds, and those agreements that impose long-term, complex legal obligations on the United States. For example, the deal by the United States, Iran and other countries is not a Treaty.

The Supreme Court ruled in the Head Money Cases that "treaties" do not have a privileged position over Acts of Congress and can be repealed or modified (for the purposes of U.S. law) by any subsequent Act of Congress, just like with any other regular law. The Supreme Court also ruled in Reid v. Covert that any treaty provision that conflicts with the Constitution are null and void under U.S. law.

Indian Law

In India, the legislation subjects are divided into 3 lists -Union List, State List and Concurrent List . In the normal legislation process, the subjects in Union list can only be legislated upon by central legislative body called Parliament of India, for subjects in state list only respective state legislature can legislate. While for Concurrent subjects, both center and state can make laws. But to implement international treaties, Parliament can legislate on any subject overriding the general division of subject lists.

Treaties and Indigenous Peoples

Treaties formed an important part of European colonization and, in many parts of the world, Europeans attempted to legitimize their sovereignty by signing treaties with indigenous peoples. In most cases these treaties were in extremely disadvantageous terms to the native people, who often did not appreciate the implications of what they were signing.

In some rare cases, such as with Ethiopia and Qing Dynasty China, the local governments were able to use the treaties to at least mitigate the impact of European colonization. This involved learning the intricacies of European diplomatic customs and then using the treaties to prevent a power from overstepping their agreement or by playing different powers against each other.

In other cases, such as New Zealand and Canada, treaties allowed native peoples to maintain a minimum amount of autonomy. In the case of indigenous Australians, unlike with the Māori of New Zealand, no treaty was ever entered into with the indigenous peoples entitling the Europeans to land ownership, under the doctrine of *terra nullius* (later overturned by *Mabo v Queensland*, establishing the concept of native title well after colonization was already a *fait accompli*). Such treaties between colonizers and indigenous peoples are an important part of political discourse in the late 20th and early 21st century, the treaties being discussed have international standing as has been stated in a treaty study by the UN.

Prior to 1871, the government of the United States regularly entered into treaties with Native Americans but the Indian Appropriations Act of March 3, 1871 (ch. 120, 16 Stat. 563) had a rider (25 U.S.C. § 71) attached that effectively ended the President's treaty making by providing that no

Indian nation or tribe shall be acknowledged as an independent nation, tribe, or power with whom the United States may contract by treaty. The federal government continued to provide similar contractual relations with the Indian tribes after 1871 by agreements, statutes, and executive orders.

Marine Life Protection Act

The Marine Life Protection Act (MLPA) was passed in 1999 and is part of the California Fish and Game Code. The MLPA requires California to reevaluate all existing marine protected areas (MPAs) and potentially design new MPAs that together function as a statewide network. The MLPA has clear guidance associated with the development of this MPA network. MPAs are developed on a regional basis with MLPA and MPA specific goals in mind, and are evaluated over time to assess their effectiveness for meeting these goals.

Overview

Unlike terrestrial conservation, marine conservation often lacks a systematic approach to conserving biodiversity. Little gap analysis has been performed on the marine environment, and there is a lack of knowledge into what is protected, what needs to be protected, and where the protection needs to occur. Over the last century there has been a rapid increase in the loss of marine biodiversity and habitat degradation. About 70% of California's population lives within one hour of the coast and the ocean provides resources to local, state, and national interests. As a result, species and habitat loss has become a major issue. Over 90% of California's coastal wetlands have been lost, coastal waters have become contaminated with a variety of urban and agricultural toxins, and a large number of targeted species have declined in the last 10–20 years. Over the last two decades, California fish catches have decreased by over 50%. These impacts have decreased the health and value of the California's coastal ocean and imply a need for a more systematic approach to marine conservation. Although there is no single solution to conserving the marine environment, MPAs are a potentially valuable tool for marine conservation when designed and managed effectively. A well designed and managed network of MPAs helps to prevent degradation, fosters marine biodiversity, and may maintain a more sustainable fishing industry. The MLPA helps to promote a shift from single-species management to an ecosystem based management and is a more systematic approach to marine conservation.

A Brief History of California Mpas

California's first six MPAs were created between 1909 and 1913; by 1950 all had been removed. After 1950 more than 50 other MPAs were created along the California coast. But these MPAs were established in a random manner and without regard to regional conservation goals. Most have been thought to be too small and ineffective in protecting against habitat and species loss. With these existing MPAs less than 1% of coastal waters were protected, and none extended to deeper waters. In 1999 the MLPA was created in order to re-evaluate the current MPA system and to establish a better network of MPAs that would be more effective in protecting against habitat and species loss.

The Marine Life Protection Act

The Marine Life Protection Act language as amended to July 2004

MLPA Findings

The MLPA found that existing MPAs were not created under a coherent plan or scientific guidelines, and that there is a need to redesign the MPA system. Coastal development, water pollution, and other human activities are a threat to California's diverse coastal waters. These coastal waters, along with the ecosystems and species which thrive within them are vital assets to the state and nation. An improved MPA system would help protect against habitat and ecosystem loss, conserve biological diversity, provide safe breeding grounds for fish and other marine species, improve research opportunities, create a reference point from which the rest of the ocean can be compared against, and may help to re-grow depleted fisheries.

MPA Network

The MLPA appointed the California Department of Fish and Game (CDFG) with the task of developing and managing a network of MPAs. The CDFG determines the final location and size of each MPA. The goal is to establish a network of MPAs that work together. This network takes into account the movement of adult and larval fish and also focuses on deepwater habitats for the first time. A proportion of the MPA network is to be designated as no-take zones. No-take zones allow for a large area of safe breeding grounds and a sanctuary for large, female fish. Large female fish produce more viable offspring and are vital in a population. With this idea, the MPA network has the potential to boost fish populations in areas out side of MPAs. Fishery growth has been successful along the Great Barrier Reef Marine Park and the Florida Keys National Marine Sanctuary after reserves were established in these areas. The final decision of the size and location of the MPAs depends on the species and habitats effected, stakeholder and conservation goals, and how each individual MPA will function on its own and as part of the network.

MLPA Implementation

After its passage in 1999, the CDFG began to implement the MLPA. The first attempt involved a Master Plan Team which included primarily scientific experts and governmental agencies, with little input from local stakeholders. This plan failed once it was brought to the public for approval, mostly because stakeholders and other members of the public were excluded from the process. Commercial and recreational fishers showed the most resistance, stating that MPAs produce no benefits for fisheries and objecting to the size and location of the proposed MPAs. In 2002, the CDFG implemented the MLPA for a second time. This plan involved members from the Master Plan Team, as well as seven Working Stakeholder Groups, which included governmental agency officials, recreational and commercial fishing interests, recreational divers, ocean vessel representatives, environmental interests, charter boat operators, harbormasters, and scientists/educators. This attempt was more successful and gained public support, but the project lost funding in 2003 due to a poor fiscal year.

In 2004 the CDFG gained new funding from several organizations to initiate the Marine Life Protection Act Initiative. The Initiative divided the coast into sequential regions and assembled a Blue

Ribbon Task Force on Marine Protected Areas, Science Advisory Team, and Regional Stakeholder Group to develop and evaluate the first set of MPAs in the Central Coast region. On April 13, 2007, after nearly three years of public meetings and proposal reviews, the Fish and Game Commission evaluated and voted on a final MPA proposal for the Central California coast. The commission voted on a plan to establish 29 MPAs covering approximately 204 square miles (18%) of state waters with 85 square miles (8%) designated as no-take state marine reserves. The network ranges from Pigeon Point in San Mateo County south to Point Conception in Santa Barbara County, and contains several types of MPAs with varying degrees of protection.

The Central Coast plan has received high marks for scientific effectiveness. Local stakeholders developed a balanced network that protects the region's best habitat, including parts of the Big Sur Coast and Monterey Bay, while allowing continued access to most recreational and commercial fishing grounds. California's Central Coast MPAs went into effect in September 2007 and scientific baseline data has been collected over the last two years.

The Central Coast was the first to be designated, first to be violated and first to be successfully prosecuted. The SMR in Morro Bay Estuary is the recipient of two creeks that drain 75 sq miles of the watershed, known as the estero Bay Hydrologic Unit. One of these creeks (Chorro)hosts effluent from a habitual polluter upstream, the CMC (California Men's Colony State Prison). The SMR took effect in Sept 2007, and in Jan 2008, the CMC spilled sewage into Chorro Creek, which meandered into the State Marine Reserve portion of the Morro Bay Estuary. Working together, coastal activist Joey Racano, then-Gov Arnold Swarzenegger and the Central Coast Regional Water Quality Control Board together successfully prosecuted the SMR violation by getting the ACL (Administrative Civil Liability) to reflect the violation was more than just another spill, but was a spill into an MPA (Marine Protected Area) into a no-take SMR. This set the precedent for all future violators, and for upstream violators (or violations from outside the SMR boundaries) as well as a precedent set for too much chlorine in the discharge. Due to the MPA violation, the CCRWQCB staff ordered the CMC to receive a higher fine than the original ACL had specified.

The North Central Coast plan, adopted by the Fish and Game Commission on August 5, 2009, also represented a compromise between different interest groups, and protected iconic sites like the Farallon Islands, Point Reyes Headlands, and Bodega Head while leaving nearly 90 percent of coastal waters open for fishing. Regulations for the North Central Coast MPAs, which extend from Alder Creek, near Pt. Arena in Mendocino County, to Pigeon Point in San Mateo County, went into effect on May 1, 2010. The regulations established 21 marine protected areas (MPAs), three State Marine Recreational Management Areas, and six special closures, in total covering approximately 153 square miles (20.1%) of state waters in the north central coast study region. Approximately 86 square miles (11%) of the 153 square miles (400 km^2) are designated as "no take" state marine reserves, while different take allowances providing varying levels of protection are designated for the rest.

In 2008, South Coast Regional Stakeholders began a public planning process to design the part of the statewide MPA network that spans from Pt. Conception in Santa Barbara county to the U.S. border with Mexico. On Dec. 15, 2010 the CA Fish and Game Commission adopted regulations to create 36 new MPAs encompassing approximately 187 square miles (8 percent) of state waters in the study region. Approximately 116 square miles (4.9 percent) have been designated as no-take state marine reserves (82.5 square miles/3.5 percent) and no-take state marine conservation areas

(33.5 square miles/1.4 percent), with the remainder designated as state marine conservation areas with different take allowances and varying levels of protection. Implementation of the new South Coast MPAs took place on January 1, 2012.

The North Coast region, which stretches from Point Arena to the Oregon border, concluded the stakeholder planning process in August 2010. Stakeholders developed a unified proposal for their regional MPA network supported by fishermen, conservationists, and tribal representatives. The unified plan was adopted by the Fish and Game Commission on June 6, 2012; and will be implemented in 2013.

The Science

An extensive body of peer-reviewed studies on marine protected areas have concluded that well-designed networks of protected waters are effective in improving ocean health and making ocean waters more resilient. Most recently, the February 2010 issue of the Proceedings of the National Academy of Sciences (PNAS) included several new studies that showed that scientifically-based MPA networks have a net positive impact on both ecosystem productivity and associated fisheries. One of the studies found that such well-designed networks can simultaneously improve the quality of ocean habitat, increase the size and abundance of sea life, and increase fishing yields and profits. Several studies have stressed the importance of location: The location of protected areas is important. In order to be effective, marine reserves must be placed in the areas where fish and shellfish feed and breed.

MLPA Controversy / Conflict of Interest Investigation

The MLPA Initiative recently began to receive negative press from fishing rights groups and individual fishermen due to the apparent conflict of interest of some MPA officials and unfair practices in the MLPA process. On May 14, 2009, Fish and Game Commission Member Jim Kellogg called for the Commission to put the MLPA process on hold due to the state budget crisis. Senate Majority Leader Dean Florez and North Coast Assemblyman Wes Chesbro requested investigation into conflicts of interest among Blue Ribbon Task Force members as well as sources of funding for the MLPA.

On May 19, 2009, the California Fair Political Practices Commission (CFPPC) disclosed that the Enforcement Division of the Fair Political Practices Commission has started a formal investigation into Fish and Game Commissioner George Michael Sutton, under charge of having violated the Political Reform Act (PRA) of 1974 due his conflicts of interest on votes on the MLPA while serving on the Fish and Game Commission.

On June 24, after investigating the matter, the Fair Political Practices Commission (FPPC) declared that Commissioner Sutton can participate in any and all public processes and deliberations surrounding the Marine Life Protection Act (MLPA) without conflict.

Lawsuits

On January 27, 2011, a number of member organizations of the Partnership for Sustainable Oceans led by Robert C. Fletcher (a former California Department of Fish and Game Chief Deputy director) filed a lawsuit challenging closures of certain fishing areas under the California Fish and Game Commission (CGFC). The plaintiff's challenged the legality of the CGFC's ability to impose its own regulation and claimed that the commission had violated the state's Administrative Procedures Act and the California Environmental Quality Act.

As part of administrative state government process, the Game and Fish Commission is required by law to disclose to the public all matters concerning MLPA research, regulation and closures. The BRTF (Blue Ribbon Task Force) and MPT (Master Plan Team), who were charged with the task of conducting research and establishing a conservation plan, had been executively ordered not to disclose the information it had obtained and produced by order of California Resource Secretary Mike Chrisman. Documents obtained in the PRA lawsuit included an email dated April 7, 2007 and advised BRTF members to "give your own notes verbally and throw them away after." This constituted a breach of the California Public Records Act, which states that "the people have the right of access to information concerning the conduct of the people's business".

On March 10, 2011, a California Superior Court ruling found in favor of the plaintiffs citing that the BRTF and MPT had failed to produce evidence that it complied with the California Public Records act while conducting research from April 2007 through November 2009. It cited that the practices used by the privately funded BRTF and MPT under the GFC did not comply with "open and transparent" processes as outlined in the Public Records Act and ordered the California Game and Fish Commission to pay all legal fees incurred by Mr. Fletcher and his team.

The Fletcher Court decision further declared that the BRTF and Master Science Team were "public bodies." According to the Court: "Based on the facts present here, they cannot be characterized as private contractors or consultants or truly independent advisory bodies, but are "Sate bodies" engaged in state governmental decision making." The Court arrived at this decision by applying the criteria of the California State University Fresno case that whether a particular entity qualifies as a governmental agency depends on whether it is performing governmental actions. The Court further found that private individuals on the BRTF and Master Plan Team were not entitled to privacy rights over public records.

There is a second interesting superior court case of Gurney v. California Department of Fish and Game et al. Superior Court of California in and For the County of Mendocino Case #SCUK-CVG-10-57448. Gurney had attempted to video tape a North Group Initiative Stakeholder Group (SG) meeting where the SG group was split in half to develop different marine reserves or arrays. He was forced to stop for disrupting the meeting. Gurney privately financed suing the Initiative and Fish and Wildlife in order to establish that the Bagley Keene Open Meeting Act rules which allow video taping, applied. While the Court could have decided the case on the narrow technical grounds that neither divided group constituted a majority of a public body and therefore there was no quorum, the court issued a more fundamental decision that the SG was not a public body so there were no rights to sue for violations of the California Public Meeting Laws. The court contrasted the SG with the clearly legislatively created and public meeting body the SAT or Master Planning Team.

There is a fascinating side decision in the Gurney case regarding the naming of Melissa Hansen-Miller an employee of the initiative as a party. The California deputy attorney argued in a motion to dismiss her as a named party of the initiative that:

The MLPA Initiative moreover is not an organization, agency, or association of any kind which may be sued in a court of law. While staffed by CNRA and DFG the "Initiative" is not incorporated. It has no officers, and it has no members or association.

The Court granted the motion to dismiss Melissa Hansen-Miller from the case on these grounds.

Despite this *de minimis* to non-existent legal status the 38 million budgeted and highly organized initiative created by Memorandum of Understanding between the Department of Fish and Game, the Resources Agency and the Resource legacy foundation conducted the largest marine planning effort in California if not U.S. history from 2004 to 2011.

There were substantial controversies between Native Americans and the North Group Science Advisory Team (SAT). The SAT initiated a survey of Native American harvesting which was conducted prior to the issuance of an Institutional Review Board (IRB) permit or conditional exemption. The Tribes wanted to keep answers confidential and have greater Native American participation in the survey. The SAT agreed to keep the answers confidential and established procedure to do so. The SAT eventually got the necessary I.R.B. permit exemption although there remained a controversy of comingling information gathered before the conditional exemption was issued. Native Americans wanted to have a qualified Native American appointed to the SAT . Their first nomination was rejected as unqualified. The second qualified nomination was not acted upon by the Director of Fish and Wildlife. The SAT was asked to treat Native American harvesting separately from state recreational harvesting and the SAT answered that was not legal. On February 11, 2010 meeting the SAT substantially changed the statewide definition of take used in all other marine regions for the Levels Of Protection (LOP) model. It was changed to be the maximum amount allowed by state and federal law. This meant that a million plus recreational license holders were assumed to harvest the full limit every single day within each proposed North Group marine reserve. Under this assumption Native American harvesting in thinly populated northern California, in a single reserve, for a single day, was greater than the annual harvest of all of Southern California. Native Americans erupted in a demonstration before the SAT at the June 29, 2010 SAT meeting demanding access to the SAT. Individual, Tribes, and North Coast Tribal Chairman's Association, followed up with requests to allow Tribal PH.D. and other Tribal scientists to get on the agenda and submit written peer reviewed marine studies. These requests were denied by the SAT. There was a second demonstration on November 17, 2010 by scientists from the Tribes who professionally objected to not being allowed to present to the SAT. The scientists stated their Ph. D. or other educational qualifications and their area of marine research. Native Americans were never allowed to present on the last minute SAT take assumption changes to the Statewide Levels of Protection Model. The LOP model was never independently peer reviewed. The amended LOP model and the science behind it remain controversial to this day.

References

- Acharya, Deepak and Shrivastava Anshu (2008): Indigenous Herbal Medicines: Tribal Formulations and Traditional Herbal Practices, Aavishkar Publishers Distributor, Jaipur- India. ISBN 978-81-7910-252-7. p. 440

- Dean, Bartholomew 2009 Urarina Society, Cosmology, and History in Peruvian Amazonia, Gainesville: University Press of Florida ISBN 978-0-8130-3378-5

- Graham A. Cosmas. MACV: The Joint Command in the Years of Escalation, 1962-1967. Government Printing Office. pp. 145–. ISBN 978-0-16-072367-4.

- Oscar Salemink (2003). The Ethnography of Vietnam's Central Highlanders: A Historical Contextualization, 1850-1990. University of Hawaii Press. pp. 28–. ISBN 978-0-8248-2579-9.

- McElwee, Pamela (2008). "7 Becoming Socialist or Becoming Kinh? Government Policies for Ethnic Minorities in the Socialist Republic of Viet Nam". In Duncan, Christopher R. Civilizing the Margins: Southeast Asian Government Policies for the Development of Minorities. Singapore: NUS Press. p. 182. ISBN 978-9971-69-418-0.

- Ranjan, Amitav (21 September 2003). "Sahib Singh wanted to visit Serbia to meet fellow Jats". The Indian Express (New Delhi). Retrieved 15 March 2016.

- "Draft Guidance for Industry: Voluntary Labeling Indicating Whether Food Has or Has Not Been Derived From Genetically Engineered Atlantic Salmon". Food. US FDA. November 19, 2015.

- "Mission to Vietnam Advocacy Day (Vietnamese-American Meet up 2013) in the U.S. Capitol. A UPR report By IOC-Campa". Chamtoday.com. 14 September 2013. Archived from the original on 22 February 2014. Retrieved 17 June 2014.

- Taylor, Philip (December 2006). "Economy in Motion: Cham Muslim Traders in the Mekong Delta" (PDF). The Asia Pacific Journal of Anthropology. The Australian National University. 7 (3): 238. doi:10.1080/14442210600965174. ISSN 1444-2213. Retrieved 3 September 2014.

- Hall, Gillette, and Harry Anthony Patrinos. Indigenous Peoples, Poverty and Human Development in Latin America. New York: Palgrave MacMillan, n.d. Google Scholar. Web. 11 Mar. 2013

- Charles Theodore Greve (1904). Centennial History of Cincinnati and Representative Citizens, Volume 1. Biographical Publishing Company. p. 35. Retrieved 2013-05-22.

Permissions

Index

www.ingramcontent.com/pod-product-compliance
Lightning Source LLC
Chambersburg PA
CBHW061256190326
41458CB00011B/3686